Biblical Mathematics

Keys to Scripture Numerics
The Significance of
Scripture Numbers as Revealed
in the Word of God

ISBN 0–937422–38–X

Published by
The Olive Press
P.O.Box 280008
Columbia, SC 29228

FOREWORD

It is the purpose of this book to give to the earnest student of the Scriptures a new field of study in the Bible. This is a study that only a few are acquainted with, but I assure you that it is an interesting one, provided you are willing to study. It will take time and effort but it is worth it. To me, this is one of the most interesting studies I have ever undertaken. Some ten years of my life have been spent in the study of Scripture Numerics. It was my study in the Book of Revelation that led me to the door of Bible Arithmetic. During this past decade I have tried to open this door wide to the hearts of God's people.

I claim no credit for any discovery of this study. Without the Holy Spirit, the Teacher of the Word of God, all our efforts would have been in vain. He is the One Who "reveals the Truth". I am indebted though to some outstanding men who have written on this subject. There are some men who have spent many years of their life in seeking the truth of the Numbers of the Word of God. To these, we express our gratitude.

My desire is, that this study will generate a deeper interest in the search of the Scriptures, and will arouse a new desire in the hearts and minds of every one who reads it; to begin anew the study of the marvelous Word of God. There is so much hidden away under the surface of the Scripture than most folk ever dreamed of. There is no other book in all the archives of the world as the Word of God.

I believe as we enter into the study of this subject, we could say with Exodus 3:5,

> "Put off thy shoes from off thy feet, for the place whereon thou standest is holy ground."

Isaiah was lifted into the throne room and saw the Lord. He heard one challenge and was prepared for and permitted to respond to that challenge. (Isaiah 6:1)

Dedication

To My Daughter — Donna Sue

My "number two" daughter, who has
been a blessing to Mother and Dad and
in a very special way has filled our
Book of Remembrance with many
happy hours,

Is This Book Lovingly Dedicated.

20th Printing, 1998

The apostle Paul was caught up into the "third heaven" and "heard unspeakable words, which it is not lawful for a man to utter" (II Corinthians 12:1-4).

Peter, James and John saw a bit of heaven on the Mount of Transfiguration, and heard Jesus talking there with Moses and Elias (Mark 9:1-13).

Stephen "looked . . . into heaven" and saw "the glory of God, and Jesus standing on the right hand of God" (Acts 7:55; cf. Isaiah 6).

I pray in this study the student of the Word will find himself standing in the very council room of God and hear the eternal God announcing His plan and program for the ages and eternity! There, in that wondrous Presence, the student will feel himself caught away from the world and from men.

God, Whose Holy Word sublime
Spans the measurements of time;
In revelation true and pure,
Maps the ways of progress sure

The world has reached a stage in history when everyone is number conscious. Every wage earner, as well as others, has a Social Security number. The men in the armed forces have their numbers, and when their dead bodies are found, they are identified by their numbers. Every life insurance policy, sick and accident, or hospital policy, has its number. When food was rationed during World War II, every ration card bore a certain number. Cars and trucks are identified and traced, both by the numbers on the license plate, and by the serial number of the motors. Football and baseball players have their numbers on the back of their uniforms by which each player can be identified. We now have Zip code numbers for our mailing, Area code numbers for long distance calling and practically our whole life is regulated by numbers. Even so has the time come for the Lord's people to distinguish by numbers between what is true and what is false; what is of God and what is of the devil.

Recently a statement was made in one of our leading magazines that 90% of all the scientists who have ever lived are living today. That more mathematics has been created since the year 1900 than in all the other centuries of history. In this fast age of space travel and scientific developments, mathematics have become basis for our civilization. Numbers have become part of our life.

If the Bible is an inspired Book as it claims to be, are not its numbers as well as its words inspired? One of its books is called "Numbers". It is not called by that name simply because it contains the numbering of the children of Israel, but because of the significance of the numbers used in it.

One out of every five Scriptures in the Bible contains a number. Numbers are important to having a working knowledge of the Bible. The numerical structure is found every-where in the Scriptures. But what is meant by "numerical structure"? It is this: That is, five divisions of Moses' whole work, this five fold division has a meaning intimately connected with the subjects of the books themselves.

Genesis, which stands first among the Books, has for its special line of truth what would be suggested by the number ONE; Exodus, similarly, a line of truth connected with the number TWO; Leviticus, with number THREE; Numbers, with FOUR; and Deuteronomy, with FIVE.

To take of these, perhaps the simplest, the number FOUR, stands as the number of the world, and the symbol for "weakness" (which may come out in failure); and so the book of Numbers will be found to be characterized by these thoughts. It is, in fact, the testing and failure of Israel in the wilderness -- the type of our own path in the world; and the characters implied in the number are found in it throughout.

Some might say that this numerical system is too artifical -- too mechanical -- seems to make the interpretation of Scriptures too independent of the Spirit of God to be of Him. The perfectly sufficient answer to this would be that it is there; and being there, it must be of Him. God's ways are often strange to us, and we misjudge strangely. Who would have thought that the alphabetic psalms would be worthy of the

Spirit of God to write? Probably no one, if He not confessedly done so. And these alphabetic psalms are but the indication of that very numerical structure which in the hundred and nineteenth psalm stamps it everywhere with the number EIGHT, which reveals easily one of its main features.

The fact remains; the numbers are there. Let criticism do its work thoroughly, and prove if they are not. Let it be as severe as the subject demands, and let the pretension be exposed, if it be merely that. Certainly it ought to be easily disproved if untrue, for never did a system submit itself to more rigorous tests than does the present one.

INTRODUCTION

In this book you are about to read, much emphasis will be placed on the spiritual significance of numbers and their perfect relationship to the Scriptures. This consistency that is found in the Bible and all its teachings will provide an inspiration helping us to follow God and His direct plan for our lives.

Numbers occupy a very important place in my personal life. Fifty-eight is the number of years I have been allowed to live on God's earth and enjoy His blessings. Thirty-six represents the number of years I have been married to my precious wife, Doris. Thirty-eight is the number of years I have been saved and serving the Lord. Thirty is the number of years I have been privileged to serve as pastor at Glen Haven Baptist Church. For eight years I have been acquainted with evangelist, author, Bible teacher Ed Vallowe, and six is the number of years he has been our Jubilee Bible teacher in the morning services. In all of my years of ministry I have been allowed to meet and be with some of God's choice servants and preachers. Among those, Brother Ed Vallowe stands out very vividly in my mind and heart for the type of man that he is. My first association with him was with his teaching of the Book of Revelation. Even though this is his specialty, I was soon to find out that he also was very proficient in many other areas such as the Book of Daniel, John, Prophecy, and even other subjects on biology, geology, botany, science, space, UFO's, and an endless list too numerous to mention in this introduction. For a man to have so much knowledge about the Bible and secular subjects and still maintain a loving, humble attitude, I find Brother Ed Vallowe to be one among many and I count it a privilege to be acquainted with him and his ministry and consider it an honor to be his pastor. Our church has been able to maintain a tremendous relationship with this man of God for several years and, our love, respect and appreciation for him grow more and more each day.

You will benefit not only from this book by Brother Ed but also from all the books he has written. The hours that God's man has set aside for prayer, meditation and study will enable all of us to see many beautiful truths and promises in God's Holy Word. Brother Ed's books are written to be read and the simplest of souls, as well as those who have studied for many years, will find them challenging, interesting and informative.

This book is written just as if Brother Ed were speaking to you personally and sharing the truths and many gems that have been revealed to him by the Holy Spirit through God's Eternal Word.

Rev. Randy Hardeman, Pastor
Glen Haven Baptist Church
Decatur, Georgia 30032
January 1977

PREFACE

In many revivals conducted throughout our nation I have been asked by multitudes of God's people to place in written form the Significance of Numbers as revealed in the Word of God.

Due to the full time itinerary of Bible conferences and evangelistic meetings, I have been unable to publish such a treatise until now. With the help of some kind friends, this work is now made possible.

It has been said that "There is no language without some numerals". In every race, tribe, and language of man, numbers have had a definite place in the life of its people. In the earliest form of life, man developed a method of counting by using his fingers. It is certain that a system of mathmetics was developed from the ancient Egyptians and Babylonians. The Arameans and Phoenicians used an upright line for 1, two such lines for 2, three for 3, and so on, and special signs for 10, 20, and 100.

It had been conjectured that these or similar signs were known to the Hebrews of the Old Testament days, but acutal proof was not forth-coming until the discovery of Jewish papyri at Assuan and Elephantine in 1904 and 1907. In these texts, ranging from 494 to 400 B.C., the dates are stated, not in words, but in figures of the kind described. We have therefore clear evidence that numerical signs were used by members of the Hebrew people in Egypt in the 5th century B.C.

The system of counting followed by the Hebrews and the Semites generally was the decimal system, which seems to have been suggested by the use of ten fingers. In the Hebrew language they have no separate symbols for numbers, corresponding to our modern Arabic figures 1, 2, 3, 4, 5, 6, 7, 8, 9, 0. In their place they make use of the letters of their alphebet; so that each Hebrew letter stands also for a certain number; and this is called the Numeric value of the letter. This is also true of the Greek language in which the New Testament was written. By means of these numeric values the Greek and Hebrews performed all their Numeric operations. (more of this is explained in the last chapter of this book).

Of the sexagesimal system, which seems to have been introduced into Babylonia by the Sumerians and which, through its development there, has influenced the measurement of time and space in the western world even to the present day.

When the Hebrews were in captivitiy in Egypt for a period of some 400 years, the Lord raised up a leader to deliver His people from bondage. Moses, trained in the finest universities of his day, was selected and prepared by the Lord God to bring His people once again into a land given them through the seed of Abraham, Isaac, and Jacob.

With this keen background God used Moses to write the first five books of the Old Testament and in them He placed the secret code of God's message to His people. This special message was hidden in the numbers of the Books. And only those who searched the Scriptures would be able to find this "manna" as they decipher the message of the numbers. Moses had more to say about the numbers of the Bible than any writer except the Apostle John.

This treatise is issued with the prayer that the Lord may abundantly bless the contents of its pages, that many may be led to more earnestly read and study the "Word of God", to be delivered from error to truth, and that others may come to know Christ as a Personal Saviour.

"He that hath ears to hear, let him hear". (Matthew 11:15)

Evangelist Ed F. Vallowe

TABLE OF CONTENTS

GOD'S DESIGNING HAND SEEN IN THE NUMBERS

In these last days when a turning away from God's inspired Word is so woefully evident all over the world, there can be no study more helpful and strengthening to the believer's heart than the subject of Spiritual Numerics, as revealed in God's Word. For the reason that it so clearly shows, to any but the most wilfully hardened hearts, that one Supreme Mind must have been the Author of all the books of which the Bible is composed.

It would have been a matter of absolute impossibility for men of different minds, surroundings and circumstances, and so widely separated by many hundreds of years in point of time, to have written unaided, books which one and all, exhibit the same wonderful use of numbers as a means of portraying deep spiritual truths, and which preserve uniformly the same meanings to these numbers in books of entirely differing character and subject matter.

Spiritual Numerics thus proves that there is only one solution to the problem, and that is the solution given by God's Word itself:

> "For the prophecy came not in old time by the will of man: but holy men of God spake as they were moved by the Holy Ghost." (II Peter 1:21)

The Holy Spirit was the One sole Writer of the Word of God from beginning to end, though many human pens were used in the actual transcribing of it. There is absolutely no escape to the conscientious student of Spiritual Numerics from the conclusion, that, in the original tongues in which the Scriptures were written, we have the very Words of God Himself.

How comforting a conclusion this is, in these last days of universal apostasy, when "mighty men are around us falling", we have a sure deep heart-rest and peace that the world does not and cannot understand. It assures the believer in God's Word that he is reading the very words of the Holy Spirit of God, who knows "the deep things of God", and who wrote these things "that we might know the things which are freely given to us of God."

The whole structure, then, of what is blasphemously called "Higher Criticism", falls with a crash to the ground. Built as it is, purely upon men's theories, which have no foundation in fact, it cannot stand before the majesty of God's pure Word, which reveals itself as the "Critic", which is "sharper than any two-edged sword", revealing to all those who prayerfully study it, "the thoughts and intents of the heart". It is indeed, as our Lord Himself declared, the "Judge" that shall judge the sinner "in the last day".

In pursuing our study we shall find there are certain rules and laws governing Bible Arithmetic, which will prove themselves as we advance, and which it may be well to state here, in order to get a clear grasp of the subject at the outset.

1. All the simple numbers from 1 to 40 have a spiritual meaning attached to them. Above 40 only a certain group of numbers will have a special spiritual meaning applied to them.
2. Numbers compounded of these numbers, e.g., by doubling or trebling, generally carry the same spiritual meaning only intensified.
3. Numbers compounded by adding two simple numbers together, usually carry the two meanings, of such numbers, expressed together, bringing out a deeper spiritual truth.
4. Where a compound number is divisible by several factors, it will usually be found that its spiritual truth, if any, is hidden behind its simplest factors, that is, those which are incapable of further division.
5. The first use of a number in Scripture almost invariably gives the clue to its spiritual meaning.
6. A spiritual truth does not appear to be evidenced in respect of every place where a number appears.

7. Numbers are used to convey spiritual truths in at least three ways:
   (a) By the actual use of a number.
   (b) By the number of times a special word or phrase is used by the Holy Spirit.
   (c) By the gematria or numerical value of a word or phrase.

With these guiding rules in mind, we can now proceed to our study. It should be understood that the examples given of the use of numbers in this book are very far from exhaustive, and are only intended indeed, as examples, to help devout Bible students to pursue the subject further, and set them upon the race of what will prove a mine of spiritual wealth. It will illuminate in a wonderful way portions of Scripture which may have seemed rather dry reading in past days.

We will be referring to the gematria or numerical value of words, and a short explanation of this science must first be given for the benefit of those who are ignorant of it. "GEMATRIA" means "word-measuring", and it as a matter of fact is familiar to almost everyone, though they may not have heard the term before.

The expression of numbers by figures, Gematria, is comparatively modern, and the ancients having no such symbols to express numbers, used letters of the alphabet instead.

The Romans used six of such letters only, namely, D to express 500, C to express 100, L to express 50, X to express 10, V to express 5, and I to express 1. (The letter M often now used to express 1000 was not used by the Romans, but is the outcome of the symbols which they used for this number, being eventually written as M. These symbols were C I which written rapidly and joined together look very much like an M.)

Most people have been made familiar with the gematria of these letters by the fact that they are still often used in the headings to chapters, and on the foundation stones of buildings. Roman gematria expressed other numbers by combinations of these letters, and by the order in which they were placed. For instance, a lower number before a higher meant that it was to be subtracted from that higher number, as for instance, XL meant 50-10=40, whereas LX meant 50+10=60.

The Hebrews and Greeks, however, went further than the Romans, and used every letter of their alphabet to express a figure. That is to say, they used their alphabets for a double purpose. The letters were letters to form words, but were also used as symbols for numbers, and as such they were used in arithmetic just as we use our Arabic numerals. For instance, the letter "a" in Greek was always used to mean "1" in arithmetic, the letter "i" similarly was used to express "10", the letter "r" to express "100". Hence it comes to pass, naturally, that any word in either the Hebrew or Greek languages can be expressed in figures, by the simple process of putting against each letter, the number or figure, which that letter of the alphabet stood for, and the addition of the values of these letters gave the gematria or numerical value of the word.

These numerical values are often very significant, and there can be no doubt in the mind of any who study the subject carefully, that the Holy Spirit uses this hidden arithmetic to prove the truth of God's Word.

We give below the numeric values of the Hebrew and Greek alphabets, that is, the number which each letter of the respective alphabets stood for:

Hebrew.			Greek.		
Aleph ..	א —	1	Alpha ..	α —	1
Beth ..	ב —	2	Beta ..	β —	2
Gimel ..	ג —	3	Gamma ..	γ —	3
Daleth .	ד —	4	Delta ..	δ —	4
Hei ..	ה —	5	Epsilon ..	ε —	5
Vav ..	ו —	6	Zeta ..	ζ —	7
Zain ..	ז —	7	Eta ..	η —	8
Cheth ..	ח —	8	Theta ..	θ —	9
Teth ..	ט —	9	Iota ..	ι —	10
Yod ..	י —	10	Kappa ..	κ —	20
Chaph ..	כ —	20	Lambda ..	λ —	30

Lamed	.. ל —	30		Mu	.. μ —	40	
Mem	.. מ —	40		Nu	.. ν —	50	
Nun	.. נ —	50		Xi	.. ξ —	60	
Samech	.. ס —	60		Omicron	.. ο —	70	
Ayin	.. ע —	70		Pi	.. π —	80	
Phe	.. פ —	80		Rho	.. ρ —	100	
Tsaddi	.. צ —	90		Sigma	.. σ —	200	
Kooph	.. ק =	100		Tau	.. τ —	300	
Resh	.. ר —	200		Upsilon	.. υ —	400	
Scheen	.. ש —	300		Phi	.. φ —	500	
Tav	.. ת —	400		Chi	.. χ —	600	
				Psi	.. ψ —	700	
				Omega	.. ω —	800	

From the above can be seen how the Name given to our Lord Jesus, in Greek terms, has the numeric value of 888, as stated in the chapter on the Number eight:

I	.. .. ι	=	10
E	.. .. η	=	8
S	.. .. σ	=	200
O	.. .. ο	=	70
U	.. .. υ	=	400
S	.. .. ς	=	200
			888

Similarly, Damascus, the oldest city in the world, has a gematria or numeric value of 444, as follows:

D	.. .. ד	=	4
M	.. .. מ	=	40
Sch	.. .. ש	=	300
K	.. .. ק	=	100
			444

The significance of this number will be seen when we come to discuss the number, four.

Another well-known Greek word, "amen," usually translated "verily" in the New Testament, has the numeric significant value, 99:

$$\begin{aligned} \alpha &= 1 \\ \mu &= 40 \\ \eta &= 8 \\ \nu &= 50 \\ & \overline{} \\ & \ 99 \end{aligned}$$

Many other examples will be referred to under the several numbers, which will make the subject clear to the student. But if it is desired to have clearer understanding and more specific proof of the use of this form of arithmetic in the Word of God, the student is challenged to take the original writings of the Old and New Testament and proceed further in this study.

NUMBERS ARE THE SECRET CODE OF GOD'S WORD

Numbers are the secret code of God's Word. Only the students of the Word, those to whom God's Spirit has given spiritual insight, will the code be made plain.

God has been called "The Great Geometrician" and is said to do everything after a plan and by number, weight, and measure. If God is the Author of the Scriptures and the Creator of the Universe, and He is, then the Word of God and the Works of God should and will harmonize.

In Revelation 13:18 the Lord has revealed a clue to the meaning of numbers.

> "Here is wisdom. Let him that hath understanding COUNT the number of the beast, for it is the number of a man; and his number is six hundred three score and six."

If then, the beast is to be identified by the use of numbers, and by counting, does not this number have a significance which is stamped with the mark of divine inspiration? Wouldn't it also be common sense to expect other numbers in the Bible to have meaning also? Is one not to learn to count the Bible as well as read it?

The psalmist, speaking of the Lord, said,

> "He telleth the NUMBER of the stars; He calleth them all by their names, Great is our Lord, and of great power: His understanding is INFINITE". (Psalm 147:4-5)

Then in Isaiah 40: 25-26, God said,

> "To whom then will ye liken me, or shall I be equal? Saith the Holy One. Lift up your eyes on high, and behold who hath created these things, that bringeth out their host by

> NUMBERS; He calleth them all by names, by the greatness of His might."

Jesus said to His disciples,

> "The very hairs of your head are all NUMBERED". (Matthew 10:30).

Why are such numbers as SEVEN and TEN found so often in books of prophecy like Daniel and Revelation if they are without significance? Why is the number THREE associated so often with the resurrection of the body if it has no significance at all?

The Bible is a book from beginning to end that is build upon a vast system of numbers which is interwoven with the doctrines of the Word of God. Such a vast system cannot be gainsaid. The agreement and harmony of the different writers, from one end of the Bible to the other on the meaning of such numbers as THREE, FOUR, FIVE, SIX, SEVEN, EIGHT, NINE, TEN, ELEVEN, TWELVE, and so on, present an argument for the inspiration of the Word of God that can never be met. Somewhere the various writers of the different books, men who lived in different ages, and most of whom never saw the others, would have crossed up each other had they not all have been guided by a master mind, ONE, Who never makes a mistake, and Whose knowledge and wisdom comprehended the events of all time.

The precision with which the Bible numbers all fall into their places cannot be accounted for except by the supernatural power and wisdom of a God Who is infinite. Could this writer, or any other man, or any number of men working together, have devised such a scheme of numbers and have made them fit from one end of the Bible to the other? Let the infidel or modernist, or agnostic explain this. Before the wisdom of the Almighty God Who devised this system of numbers, the mind of man is helpless.

Numerics are found in all creation. Every snowflake is composed of ice crystals built upon a definite symmetrical plan. Seven runs through animal life in periods. After the French Revolution, wise men enacted a rest day once in ten, thinking one in seven was too wasteful. The first sign of anything wrong was shown by the donkeys, who broke down under the strain. Then it was discovered that man's pulse beats slower every seventh day! The sabbatic rest had to be restored. The reason we are having so many heart attacks, mental breakdowns among our people, is, we have neglected to observe this one principle of rest as taught in the Word of God. Is it surprising that God should put numeric features in His Book?

Just as ropes in the British navy used to be identified by the scarlet thread running through them, so American dollar bills have little red marks in them to guard against counterfeiting. Why may not God have put features in His Book to prove that it is inspired? His truth is like a jewel in a beautiful casket.

Who but God could control the birth rate and death rate and fix it so where just ELEVEN groups of people would spring from Canaan, Ham's son, upon whom a judgment, represented by the number ELEVEN, was pronounced?

How did it happen that Shem, the son of Noah, through whom Christ came, had just FIVE sons, representing the GRACE that was to come through Christ? How did it happen Arphaxad, through whom Christ came, was the THIRD son of Shem, thus representing the RESURRECTION which came through Christ?

Christ came of the tribe of Judah. (Hebrew 7:14) Grace came through Christ. (John 1:17). How did it happen that Judah had just FIVE sons?

Science and certain experiences and observations common to all men show that the Universe, is built upon mathematical principles; that the Author of Nature is a most marvellous mathematician. If the Creator, Whose glory the Heavens declare, chose to produce a Book, would it not be produced on the same principles of mathematical perfection, so that He can say "The Law of the Lord is perfect"? As no man can claim to be the Creator of the mathematical wonders of the Heavens, so God has written His Signature in His Word, that no man can claim it as a human production.

These things bring man face to face with God, the Author of the Bible, and the One before Whom those who reject His Word will some day be brought to judgment. These things challenge the infidel and agnostic to account for what could not possibly have been done by human wisdom and ingenuity. This challenges the mathematical mind of this last age to see and know the MASTER MIND and to come to RECEIVE His Only Begotten Son as their Saviour.

Not only does this system of Bible numbers establish the Inspiration of the Scriptures, but it also established the doctrine of VERBAL INSPIRATION. By verbal inspiration we mean that not only did God inspire the thoughts of the Bible, but that He also inspired men to use the very words which they wrote. There have been some among us who claimed that God only inspired the thoughts of the Bible, and that men were left to choose their own words to convey those thoughts. That theory of inspiration crumbles to dust beneath the weight and evidence of Bible numbers, and the doctrine of VERBAL, or word inspiration, stands stronger than the rockribbed hills. The numbers of the Word fall in their places because of the verbal inspiration of the Word.

First of all, I want to establish logically, in a manner that cannot be gainsaid, that the Bible is verbally inspired of God; that man could not have written it like this; that the only explanation therefore for the facts, is a Superior Mind, Who does not leave a single department of nature without His mathematical signature, and Who has also shown thus His Hand in the Bible. And the reason for this is quite natural. We are living in the latter days, when, among many, respect for the Bible is gone. Nineteen hundred years ago the case was otherwise.

Nineteen hundred years ago, when men were confronted with the question as to the truth of the Word of God, if they were at all intelligent and candid, they could not deny the evidence. For some twnety-five years 12 picked men were going up and down the earth, appointed by our Lord to be His witnesses in a world that did not believe in Him. Men naturally asked, "Who is that Jesus of Nazareth, for Whose sake these men have forsaken their all for the witnessing that He has risen from the dead?" Not only was there no gain for them from this testimony, but persecution, shame and torment even unto death was their sole expectation therefrom. And sensible folk began to examine Christianity scientifically, and they said: "What about this story of yours, that a dead man was buried and rose on the third day"?

One after another stood and gave their eye witness testimony. Peter, James, John, Thomas, and finally Paul, who was a scholar of scholars, graduated at the great University of Gamaliel. And Paul tells how he hated Jesus and persecuted His followers unto death, yet finally had to say:

> "And now I delight to be a witness that He is risen from the dead!" (Acts 26)

Any sane man in those days, who was at all honest with himself would have said: "It is a greater miracle for these men to be lying and not telling the truth, in order to be a witness to that truth than to believe that these things are impossible."

But today we have not those men in our midst. We cannot go to Peter and say, "Tell us your story". We cannot bring Thomas and Paul back. We simply have the message here in this Book, and if this Book fails us and cannot be trusted, then of all men we are the most miserable, because our faith is vain, and we are yet in our sins.

So I can conceive of the Holy Spirit foreseeing that there will come a day when the whole battle between the intellect of man and the Faith of the Lord Jesus Christ will rest on that Book, and He uncovers its trustworthiness in these latter days as never before: first of all by Bible Numerics, and then by Archaeology. Hardly a year passes but some news comes from the excavators which once more proves that every statement in the Bible can be vertified, even by the spade.

This is why I believe that Bible Numerics has been held in abeyance by the hands of God for these last days, in order to show his Hand at the proper time, so that God should not be mocked, that His Word is true, and that His Name is above every name.

Let us proceed and demonstrate in an objective way that the Bible is actually the Word of God and inspired verbally by God, not inspired in any human sense, but actually in-breathed by the Holy Spirit.

Paul said, "The LAW worketh WRATH." (Romans 4:15) The number for LAW is TEN. In the works of the flesh as given by Paul in Galations 5:19-21, the TENTH one is WRATH.

> "Now the works of the flesh are manifest, which are these; Adultery, fornication, uncleanness, lasciviousness, Idolatry, witchcraft, hatred, variance, emulation, WRATH, strife, seditions, heresies, enveyings, murders, drunkenness, revellings, and such like: of that which I tell you before, as I have also told you in time past, that they which do such things shall not inherit the kingdom of God."

Neither was this a mere coincidence. THIRTEEN is the number for rebellion and depravity. In the things Paul listed in Romans 1:29-31, the THIRTEENTH is "Haters of God." Here is seen the depraved nature of man, represented by number THIRTEEN.

John 1:17 states, "The law was given by Moses." In the list in Hebrew 11:1-23 of those who accomplished things by faith, Moses is the TENTH, and TEN is the number for LAW.

THREE is the number for the resurrection. When Jesus raised Lazarus from the dead He spoke just THREE words to him. They were: "Lazarus, come forth." (John 11:43) Was John given the exact words to write or was he left to give the thought in his own words? Surely the Spirit of God gave him the exact words to write.

The THIRD beatitude is "Blessed are the meek: for they shall inherit the earth". (Matthew 5:5) It is stated in Psalm 37:11 that "The meek shall inherit the earth; and shall delight themselves in the abundance of PEACE." This connects PEACE with the inheritance of the earth by the meek. In Matthew 5:5 their inheritance of the earth is connected with the THIRD beatitude. The THIRD thing mentioned by Paul as the fruit of the Spirit is PEACE. "The fruit of the Spirit is (1) love, (2) joy, (3) PEACE." (Galatians 5:22)

THREE is the number for the resurrection of the body. The meek will not inherit the earth until after the resurrection. That is why the THIRD beatitude is "Blessed are the meek; for they shall inherit the earth." When they are raised from the dead and inherit the earth they shall have an abundance of PEACE. How well this fits in with the word PEACE being the THIRD mentioned in Galatians 5:22.

In Matthew 19:29 our Lord said,

> "Every one that hath forsaken (1) houses, or (2) brethren, or (3) sisters, or (4) father, or (5) mother, or (6) wife, or (7) children, or (8) lands, for my name's sake shall receive an hundred fold, and shall inherit everlasting life."

Here we have the number EIGHT to picture those with the NEW BIRTH. It is the same in Mark's account, Mark 10:29-30. On the other hand Luke only mentions FIVE.

> "There is no man that hath left (1) house, or (2) parents, or (3) brethren, or (4) wife, or (5) children, for the kingdom of God's sake, who shall not receive manifold more in this present time, and in the world to come life everlasting." (Luke 18:29-30)

In this may be seen the leadership of the Spirit of God. Matthew and Mark were inspired to list the number (EIGHT) which would show the NEW BIRTH. LUKE was inspired to list the number (FIVE) which would show the GRACE OF GOD. It took the EIGHT listed by Matthew and Mark, no more, and no less, to show that the one who would forsake these things for Christ is a born again person. It took the FIVE listed by Luke, no more, and no less, to picture the GRACE OF GOD that would enable a person to forsake all for Christ. All three of the writers were given the exact words to use.

Thus the doctrine of the Verbal Inspiration of the Word is established by the numbers. The numbers fall in their places because of the Verbal Inspiration of the Word. The Arithmography of Scripture is the death warrant of the theory of the destructive critics. The numerics of the Bible are fatal to the foes of Verbal Inspiration, and are invulnerable.

Both the pre-incarnate existence of Jesus Christ, and His incarnation are proven by Bible numbers.

> "In the beginning was the (1) WORD, and the (2) WORD was with (1) GOD, and the (3) WORD was (2) GOD. The same was in the beginning with (3) GOD." (John 1:1-2).

It will be noticed that the WORD is mentioned THREE times, and that GOD is found THREE times in this quotation. Why the number THREE if not to show that there were THREE in existence in the beginning, and the Christ, the WORD, was in the beginning and was ONE of the Divine TRINITY. This shows the existence of Christ, the WORD, as ONE of the THREE persons of the Godhead. This established the pre-incarnate existence of Christ.

The FOURTH time the "WORD" is found is in John 1:14,

> "And the WORD was made flesh, and dwelt among us, (and we beheld his glory, the glory of the only begotten of the Father) full of grace and truth."

FOUR is the number for the world, the flesh, or first creation. So the first THREE times Christ is called THE WORD is in the place where He existed as one of the Trinity in the beginning. The FOURTH time it is mentioned is where the WORD became FLESH. Here there is set forth the doctrine of the incarnation of Christ. This verse connects the doctrine of His incarnation with number FOUR. Here the Creator became also the creature.

Many times in revival meetings I have been asked of parents the destination of children or babies who have died. Where are they? Did they go to hell? Are they in heaven? Were they saved? How did they get saved?

I believe that even though a child's nature is depraved he is not condemned. This is proven by Bible numbers. THIRTEEN is the number of DEPRAVITY. In Mark 7:20-23, Jesus mentions THIRTEEN things which come out of the hearts of men, and defile them. Over and over again the Bible reveals the number THIRTEEN is connected with REBELLION. In the list of things Paul mentioned in Romans 1:29-31 the THIRTEENTH is "Haters of God." In this the depraved, rebellious nature of mankind appears.

But THIRTEEN is not the number for death, but TWENTY-THREE. THIRTEEN lacks TEN of being TWENTY-THREE. The law must be added to bring the other TEN and make TWENTY-THREE for death. Paul said, "I was alive without the LAW (10) once: but when the commandment came, sin revived, and I died." (Romans 7:9)

> "For sin (13), taking occasion by the commandment (10), deceived me, and by it slew me." (Romans 7:11) (23)

So it is clearly seen that the sinful creature's nature, represented by THIRTEEN cannot bring spiritual death until the law adds TEN more, making TWENTY-THREE, the number of DEATH.

Israel sinned several times between Egypt and Sinai, where the law was given. But there is no record of God sending death or a plague upon them from Egypt to Sinai. It takes the LAW plus SIN to bring DEATH. This is why a child does not go to hell. Until there is a realization of rebellion against God and the condemnation of the Law, salvation is not needed.

Surely no one can study this system of numbers and question the fact that God knew all things from the beginning. The God who could devise so vast a system of numbers and make it fit in so perfectly in His Word from beginning to end is unlimited. The numbers fit in the births, lives, and deaths of Bible characters in such a way as to unfold God's pattern through the ages. To fix the numbers in such a way God had to be infinite in power and wisdom. There had to be exactly sixty-six of Jacob's descendants to go down into Egypt with him to fore-shadow the future idol worship of the nation. (Genesis 46:26; Jeremiah 25:4-11). Had not two of Judah's sons died before this, that number would have been sixty-eight, which would not have represented Israel's idol worship. Then Jacob, Joseph and Joseph's two sons, who were already in Egypt, made seventy, which number is the exact number of years, they went into Babylonian captivity because of their idol worship. (Jeremiah 25:4-11) All this had to be foreseen and worked out beforehand by the Lord. No wonder the Psalmist said,

> "Great is our Lord, and of great power: his understanding is INFINITE." (Psalm 147:5)

Another great doctrine revealed by the system of Bible numbers is the Premillennial position with regard to the THOUSAND YEARS as mentioned in Revelation 20:1-7. Let the opponents challenge this teaching and prove it wrong by Bible numbers!

The Premillennialists have always taught that the first resurrection was the bodily resurrection of the saved, which takes place at the Rapture, and that the Lord's coming, or His Revelation, would take place SEVEN years later and that the Lord's coming must take place before the reign of the THOUSAND YEARS. Their position is vindicated by the use of Bible numbers.

The expression THOUSAND YEARS is found SIX times in Revelation 20:1-7. Each time it occurs the meaning of the numbers from ONE to SIX is fitting.

The FIRST time the THOUSAND YEARS is mentioned is in Revelation 20:2,

> "And I saw an angel come down from heaven, having the key to the bottomless pit, and a great chain in his hand. And he laid hold on the dragon, that old serpent, which is the Devil, and Satan, and bound him a THOUSAND YEARS."

ONE is the number for UNITY. Men have been working for unity, harmony, and peace among the nations of the earth and the people of the earth for generations, and they have failed. There is a good reason why they have failed. The DEVIL is not yet bound. Not until he is bound can there be UNITY, harmony and peace on earth. The Devil will be bound so that UNITY can be brought to pass.

The SECOND time the THOUSAND YEARS is found is in Revelation 20:3,

> "And cast him into the bottomless pit, and shut him up, and set a seal upon him, that he should deceive the nations no more, till the THOUSAND YEARS should be fulfilled: and after that he must be loosed a little season."

TWO is the number for DIVISION. When the Devil is shut up and sealed that will put an end to his work of DIVISION until the thousand years are over.

The THIRD time the THOUSAND YEARS is found is in the 4th verse,

> "And I saw thrones, and they sat upon them, and judgment was given unto them: AND I saw the souls of them that were beheaded for the witness of Jesus and for the word of God, and which had not worshipped the beast, neither his image, neither had received his mark upon their forehead, or in their hands, AND They (both groups mentioned above)

> lived and reigned with Christ a THOUSAND YEARS." (Revelation 20:4)

The writer has divided this verse so the reader can see that the martyred saints are not the only ones who will live and reign with Christ a thousand years. Opponents of the Premillennial position always ignore the first group in this verse.

> "I saw thrones, and they sat upon them, and judgment was given to them."

This is the first group. The conjunction "AND", which follows this statement, adds the martyrs of the Tribulation Period, that is, those who will be killed by the beast, to the first group in this verse. It is unfair to take the second group only and ignore the group in the first statement. But that is the method used to evade the Premillennial position.

THREE has been found to be the number connected with the bodily resurrection of Christ and His people. This is the THIRD time John mentions the THOUSAND YEARS. In the very next verse John refers back to this and calls it the FIRST RESURRECTION.

> "This is the first resurrection." (Verse 5)

The opponents of the Premillennial position have tried to insist that the first resurrection is the new birth. Here is where the Bible system of numbers blast their theory to pieces. EIGHT, and not THREE, is the number for the new birth. The first resurrection is found connected with the THOUSAND YEARS the THIRD time John mentions this time period. THREE is definitely the number that is connected with the bodily resurrection of the saved dead.

There is still a further meaning to be found in these TWO groups. But time and space forbids going into that in this place.

The FOURTH time the THOUSAND YEARS occurs is where it is said,

> But the rest of the dead lived not again until the THOUSAND YEARS were finished."

They are the unsaved dead. FOUR is the number for the unsaved, or fleshy man. Those who die in that state will not be in the FIRST RESURRECTION. They will not be raised until the THOUSAND YEARS are over.

The FIFTH time the THOUSAND YEARS is found is in Revelation 20:6,

> "Blessed and holy is he that hath part in the first resurrection: on such the second death hath no power, but they shall be priests of God and of Christ, and shall reign with him a THOUSAND YEARS."

FIVE is the number for GRACE. Those who received grace will reign with Christ.

> "Much more they which receive the abundance of GRACE and the gift of righteousness shall REIGN in life by one, JESUS CHRIST." (Romans 5:17)

That the saints do not reign in this present time is made plain by Paul when he said,

> "I would to God ye did reign, that we also might reign with you." (I Cor. 4:8)

If Paul did not count himself and his brethren to be reigning during their life time, then how dare others claim that they are now reigning with Christ? Those who receive Grace are to reign in life by Christ, and with Christ, but not until after the resurrection of the saved has taken place. We cannot now live a THOUSAND YEARS in our bodies. But we can live that long, and longer, in our glorified bodies.

There is a FIVE-FOLD division in Revelation 20:6. Here is GRACE in the reign, and in the FIFTH time the THOUSAND YEARS is mentioned.

1. Grace makes people blessed and holy.

2. Grace will give one a part in the first resurrection.
3. Grace will save from the second death.
4. Grace makes priests unto God and Christ.
5. Grace will give a part in the thousand years reign.

The SIXTH time the THOUSAND YEARS occurs is in Revelation 20:7,

> "And when the THOUSAND YEARS are expired, SATAN shall be loosed out of his prison, and shall go out to deceive the nations in the four quarters of the earth."

Here Satan's work comes in again under the number SIX. This is the SIXTH time the THOUSAND YEARS is mentioned by John. Since SIX is the number connected with Satan his work appears again in this connection. You will also note that an angel will do SIX things to Satan in Revelation 20:2-3,

1. Lay hold on him.
   (This proves that angels can actually lay hold of the bodies of each other and can be confined to material places in material chains and with doors and locks.)
2. Bind him with a literal chain.
3. Cast him into the abyss.
4. Shut him up in prison.
5. Set a seal upon him for 1000 years.
6. Loose him for a little season.

Thus the reader can see the precision with which the Bible numbers fall in their places to teach their lesson.

One more thing needs to be mentioned in this connection. The FIRST RESURRECTION is mentioned twice in Revelation 20:5-6. TWO is the number that represents DIVISION. So the first resurrection will be a resurrection that will divide the dead. It will DIVIDE the saved from the unsaved dead, when the CHILDREN OF GOD (Luke 20:35-36) shall be raised FROM the dead and separated from the unsaved dead.

This writer has been unable to find the resurrection of the unsaved dead connected with the number THREE. But it has been shown that it is connected with the resurrection of the saved. It has been shown that TWENTY-THREE is the number for death, and TWENTY is for redemption. The number THREE is connected with the resurrection of the saved. THREE from TWENTY-THREE leaves TWENTY, for the redemption of the body of the saved. If number THREE is also applied to the resurrection of the wicked then when they are raised we would have to subtract this number from TWENTY-THREE, which would leave TWENTY, and would also give to the unsaved the redemption of their bodies. Is there any hint in the Scriptures that there is redemption for the bodies of the unsaved?

The resurrection of the unsaved is referred to in four places in the New Testament. (John 5:29; Acts 24:15; Revelation 20:5, and Revelation 20:12-13). The THIRD time it is referred to is in Revelation 20:5 where the expression THOUSAND YEARS occurs the FOURTH time. In this place it distinctly says,

> "The rest of the dead lived not again until the thousand years were finished?"

This disconnects the resurrection of the wicked from the number THREE. Their resurrection is a separate resurrection, both in nature and point of time. No Scripture can be quoted that states that they will be raised when our Lord returns. Those who teach that they will are without a quotation of Scripture which proves their claim. For lack of Scripture they substitute human reasoning and deduction.

We have abundantly connected Bible numbers with the teaching of the Premillennialists. The numbers are there for the other man if he can use them. If he cannot use them there must be something wrong with his doctrine. It would be interesting to see the general resurrectionist try to apply the numbers to his theory about the resurrection? So that false doctrine, started by Augustine years ago, by which he hoped to escape the argument of two separate bodily resurrections, is forever exploded with Bible numbers.

What will Postmillennialists, Amillennialists, and Anti-millennialists say to this? They can easily see that these numbers bring the infidels face to face with an argument they can never answer. They can never account for this vast system of numbers except by admitting that the Bible was inspired of God. Here we have before us a system of Bible numbers running from Genesis to Revelation. They fit in the doctrine of the incarnation of Jesus. They fit into grace, redemption, the new birth, the security of the believer, the law, depravity, and other doctrines. They refute the infidel's claim that the Bible is just a manmade book. Why is the Postmillennialists, Amillennialists, and general resurrectionists theory of the resurrection and the thousand years reign so out of harmony with the Bible system of numbers? How is it that this writer has been able to fit in with the Premillenial position so harmoniously unless God fixed it that way?

When this writer undertook the work of writing this book it was his hope that he would put out something that no modernist or infidel could ever answer. Now he has been made to feel that this hope has been realized, even beyond what he expected when this work was begun. It has required many hours of study and work, both day and night, but the writer has been repaid many times over in his own life. His appreciation of the INFINITE wisdom of God and of His marvelous Word has been broadened and deepened.

He has always believed and taught that God is all wise. But never before has he been made to so bow before God in such wonder and amazement. How could God so arrange such a book as the Bible and make such seemingly insignificant events and statements fit into the great plan of the Book as to set forth the glorious plan of redemption is something far beyond the ability of the finite creature to comprehend. This writer knows that in himself he could never have brought to light these hidden treasures that are concealed in the Divine Word of Truth.

There is one thing that makes all this solemn yet joyful to the saint: the assurance that it seems to give that the end is nigh at hand. The very power of demonstration that is in this numerical system seems to make it as a closing testimony, -- faith almost coming to an end, -- God coming face to face with man. Here it becomes us not to go too far in assumption. His ways are not as our way. But in any case, the end cannot be far.

Let us now examine the important numbers of the Word of God and find the message revealed in each number that will open our understanding and bless us with the deep truths of God's Word.

ONE
UNITY

ONE is the number in the Bible that stands for UNITY. UNITY is an important Bible doctrine. It symbolizes the UNITY OF GOD. It stood for that which was unique and alone. Being the prime number, this naturally signifies beginning. Hence, ONE stands in Scripture for source, unity, sovereignty, creation, and chiefly God (the first, or creative, PERSON OF THE TRINITY). It comes from the word "unit".

> "There is ONE BODY, and ONE Spirit, even as ye are called in ONE hope of your calling; ONE Lord, ONE faith, ONE baptism, ONE God and Father of all, who is above all, and through all, and in you all." (Ephesians 4:4-6)

Jesus also said in John 10:30,

> "I and My Father are ONE."

Here we see the UNITY that is found in the Godhead. This truth again is portrayed in I John 5:7 when John said,

> "For there are three that bear record in heaven, the Father, the Word, and the Holy Ghost, and these three are ONE."

Jesus also prayed in John 17:21-22,

> "That they all may be ONE; as Thou, Father, are in Me, and I in Thee, that they may be ONE in Us . . . and the glory which Thou gavest me I have given them; that they may be ONE even as we are ONE."

For what else could Jesus have been praying except

for the UNITY of His people when He prayed that they should be ONE?

In Ephesians 4:1-6 the word UNITY is connected with the number ONE seven times over.

> "I therefore, the prisoner of the Lord, beseech you that you may walk worthy of the vocation wherewith ye are called, with all lowliness, and meekness, with longsuffering, forbearing one another in love; endeavoring to keep the UNITY of the Spirit, in the bond of peace. There is ONE body, and ONE Spirit, even as ye are called in One hope of your calling; ONE Lord, ONE faith, ONE, baptism, ONE God and Father of all."

There was a time when the Jews and Gentiles were separate, having little dealings with one another, and at enmity with one another. But in Christ this division is abolished and they are made ONE.

> "For He is our peace, who hath made both ONE, and hath broken down the middle wall of partition between us, having abolished in flesh the enmity, even the law of commandments contained in ordinances; for to make in Himself of the twain (Jew and Gentile) ONE new man, so making peace."

In Ephesians 4:3 Paul admonished them to keep the UNITY of the Spirit, in the bond of peace. The grace of God abolished this division that was between the Jew and Gentile and brought about the UNITY of the Spirit in the bond of peace for those who believe. This was done when the two were made ONE in Christ.

It takes more than union to bring UNITY. Men have brought about a union of nations in the hope of bringing peace on the earth. But they have not succeeded in bringing UNITY (Oneness) and peace. We can have, and do have, union and strife and war. But where there is UNITY there is peace.

In John the 10th chapter, Jesus said,

> "Other sheep I have which are not of this fold: them also I must bring, and they shall hear My voice; and there shall be ONE flock, and ONE shepherd." (John 10:16)

These words of our Lord conveyed a remarkable truth, and because the Jews were not prepared to accept this invitation to come into vital union with Christ, there was a division among them.

Another very interesting fact in connection with this number, is that in Hebrew there are two words for ONE. "ECHAD" meaning a compound or collective UNITY, such for instance as, ONE crowd, ONE flock, ONE group, etc., and "YACHEED", meaning absolute UNITY or UNIQUENESS. While the former word is used many times in the Scriptures, the latter word is used only TWELVE times, and seems always to refer in type to the Lord Jesus, the Only ONE. The first time this word occurs is in Genesis 22:2,

> "Take now thy son, thine only son, Isaac, whom thou lovest --"

Here we have the first clear type of Him Who is the Only Begotten Son.

In Psalm 22:20, the Crucifixion Psalm, this word is found again. Here Jesus is called "My Darling" (literal meaning MY ONLY ONE)

The last place this word is found in the Old Testament is in Zechariah 12:10 which states that the Jew "will mourn as one mourneth for his only ONE."

> "They shall look upon Me Whom they have pierced, and they shall mourn for Him as one mourneth for his only ONE."

Here is clear evidence and reference to the Lord Jesus, the only hope for poor-hell-deserving sinners.

These scriptures reveal two truths.

1. That the Lord God had only ONE Son whom He was giving for the life of the world.

2. There can only be true UNITY in Jewish national life, and union with God when they recognize their Messiah as the ONE whom they have rejected.

In Deuteronomy 6:4, the other word, "echad" is used for "ONE Lord", showing that God is a composite Unity. Three Persons in ONE God, thus early in Scripture giving proof of the great doctrine of the Trinity.

Then the glorious Millennium so rapidly drawing near, is singled out by the Holy Spirit to show, that then, there will be UNITY of rule, Zechariah 14:9 stating,

> "And the Lord shall be King over all the earth; in that day shall there be ONE Lord, and His Name ONE."

And it will happen on a special day, as verse 7 tells us:

> "But it shall be ONE day, which shall be known to the Lord, not day nor night: but it shall come to pass that at evening time it shall be light."

Thus, the beginning of that glorious period of UNITY of rule under ONE King is marked by ONE day, "not day nor night" like the FIRST Creation day, but on that day at evening time, LIGHT. In this wonderful way is the great lesson of number ONE, given in the FIRST chapter of the Bible, carried through from the Creation to the Millennium -- UNITY and LIGHT.

When we come to Revelation, that great Apocalypse that leads up to the consummation scenes, and ends with the vision of Eternity, we find our Lord describing Himself in the first chapter as well as in the last chapter, "The First and the Last", and again, in the letters to the churches, He censures the Church of Ephesus in His first letter, because "thou hast left thy FIRST love." This was the sin that led to an ever-widening gap between what the Church of Christ should be and what it is today. Christ is "the FIRST", and He must come FIRST in the hearts and lives of all His redeemed. He is the LIGHT, and only as we walk in the LIGHT, shall we have fellowship one with another, and thus manifest that UNITY which was so marked a feature of Pentecostal days, when "the multitude of them that believed were of ONE heart and ONE soul."

> "But if we walk in the light, as he is in the light, we have fellowship ONE with another, and the blood of Jesus Christ his Son cleanseth us from all sin." (I John 1:7)

The book of Genesis affords us another illustration of this PRIMARY number. In Genesis we see the Divine Supremacy and Sovereignty of God. God is Sovereign in creation, in giving life, and sustaining life. The name of which God especially reveals himself to the patriarchs was "EL-SHADDAI", which means, "God Almighty."

> "And I appeared unto Abraham, unto Isaac, and unto Jacob, by the name of God Almighty, but by my name JEHOVAH was I not known to them." (Exodus 6:3)

This title occurs in Genesis SIX times, and in the rest of the Pentateuch THREE times -- NINE times in all (3 x 3) the square of THREE, the number of DIVINE PERFECTION or DIVINE COMPLETENESS. All through the book of Genesis we see the Divine Sovereignty of God. See His Sovereignty in calling Abraham and NO other. (Acts 7:2); choosing Isaac and NOT Ishmael (Genesis 17:18-21); Jacob and NOT Esau; Ephraim and NOT Manessah.

This first book is the ONE book. It contains all the other books in embryo, and has been well called, "the SEED PLOT of the Bible." Its Divine title is "THE BEGINNING" the "FIRST".

> "In the beginning God." (Genesis 1:1)

GOD FIRST! Here is the beginning of life, the beginning

of prophecy. (Genesis 3:15) The woman's SEED foretold, the Serpent revealed, and Salvation provided.

The COVENANT made with Abraham (Genesis 15) was UNCONDITIONAL, because there was only ONE contracting party. The Law has a MEDIATOR, therefore there were TWO parties to that Covenant.

> "But a Mediator is not of ONE, but God is ONE." (Galatians 3:20)

God alone made this Covenant, hence it is called "the covenant of PROMISE."

The first time a word occurs in the Scripture, there is always important significance found therein. The ancient Jewish commentators called special attention to them, and laid great stress upon them as always having some significance. It generally helps us to fix the meaning of the word wherever found in the Scriptures. Take for example the word -- "Hallelujah". It occurs FIRST in Psalms 104:35.

> "Let the sinner be consumed out of the earth, and let the wicked be no more. Bless thou the Lord, O my soul, Hallelujah."

The FIRST occurrence in the New Testament will be found in Revelation 19:1-3.

> "I heard a great voice of much people in heaven, saying, Alleluia; salvation, glory, honor, and power unto the Lord our God; for true and righteous are His judgments; for He hath judged the great whore, which did corrupt the earth with her fornication, and hath avenged the blood of His servants at her hand, and again they said, Alleluia."

In both these places, in the Old Testament and in the New, the FIRST occurrence of the word "Hallelujah" stands in connection with JUDGMENT. It is a shout of VICTORY over the enemy.

There are also many words that occur only ONCE in the scriptures, which are deeply interesting to study.

Here we have only room for a few examples. In Daniel 8:13 the word "PALMONI", translated "THAT CERTAIN SAINT", but really means "the Numberer of Secrets," or "the wonderful Numberer" thus showing us at the outset of our study Who it is that has hidden this marvellous arithmetic in the Word of God.

In Matthew 11:29 the word "PRAOS" (MEEK), "LEARN OF ME; for I am MEEK and lowly in heart," shows us from Whom alone we can learn the secret of meekness, and obtain that "meek and quiet spirit, which is in the sight of God of great price."

In II Timothy 3:16, "THE OPNEUSTOS" -- meaning "GOD-BREATHED", describes the unique Scriptures of God. The only God-breathed Book in the world. "All Scripture is God-breathed."

In the life of our Lord, the FIRST words we find recorded of Him are full of significance. The Lord Jesus must have spoken often from time to time in His childhood, but the Holy Spirit was not willing to record the sayings. It was NOT until the age of TWELVE that he said unto His parents,

> "Wist ye not that I must be about my Father's business." (Luke 2:49)

These are very solemn words, especially in the light of the FIRST words thrown on the light of His LAST words, when He said, in His dying hour,

> "It is Finished." (John 19:30)

Now the question enters the mind, "What is Finished?" Jesus had reference only to ONE thing. "His Father's Business!" The Redemption and Salvation for man. When Jesus came it was wholly to accomplish the WILL of the Father. The Father's WILL is FIRST, in the life of the SON.

Here we see the PREEMINENCE of the NUMBER ONE, the UNITY with the Father. Jesus said,

> "This is the Father's WILL which hath sent me; that of all that he hath given me, I shall

lose nothing, but should raise it up again
in the last day." (John 6:39)

Thus we see that salvation was no afterthought with God, it was part of the Eternal and glorious purpose of God. Salvation FIRST, originated with and in the presence of the TRINITY. It was not just primarily for the GOOD of man, but it was for the Glory of God.

Hebrews 4:12 states that "The Word of God is a discerner (critic) of the thoughts and intents of the heart." When we rest upon the UNITY of the Book it becomes the ONE and ONLY critic. When we question its authority as The Word of God we become its critics.

In conclusion, we see that the great lessons taught by the FIRST of figures, is the FIRST things must come first. There can be no spiritual life without the FIRST thing needed, birth. And spiritual birth is only possible by receiving Him WHO IS THE LIGHT into the soul, then the FIRST day of the spiritual life begins, and through believing in Him we are brought into "the UNITY of the faith." The command is "Seek ye FIRST the Kingdom of God, and all these things shall be added unto you." (Matthew 6:33)

Always you will find UNITY when the number ONE is mentioned. There is ONE book, ONE plan of salvation, ONE faith, ONE light, ONE hope, ONE Lord, and ONE Gospel.

This number is used 1898 times throughout the Word of God.

TWO

UNION, DIVISION, WITNESSING

TWO is the number of UNION (TWO becoming ONE). This also has the meaning of DIVISION and WITNESSING. Amid the dangers of primitive life, with a fear of wild beasts, or of hostile attack by his enemies constantly before him, man gained courage in companionship. TWO were far stronger and more effective than ONE. Thus, the number "TWO" came to stand for strengthening, for confirmation, for redoubled courage and energy. As TWO were joined together in a union, strength and power was united into ONE force.

There was a symbolic significance in the fact that Jesus sent His disciples forth TWO by TWO. TWO witnesses confirmed the truth, and their testimony which otherwise would have been weak was made strong. Always this number TWO meant augmented strength, redoubled energy, confirmed power.

The Union of Marriage, and "they shall be one flesh". (Genesis 2:23, 24) Here we have man and woman, TWO, joined together by God to become ONE.

The Union of Christ and the Church in Ephesians 5:31, 32, shows us the power and strength we possess when we are joined unto Christ.

The Union of the TWO Natures in Man. When the old nature is subdued by the New Nature then we possess power and strength in the Lord.

The Union of death and life in the Atonement of Christ, as seen in the "TWO BIRDS" (Leviticus 14: 4-7), and the "TWO goats". (Leviticus 16:5-22)

There were "TWO Tables of Testimony", and "TWO Witnesses" were necessary to a fair trial, and "TWO Witnesses" will testify during the Tribulation Period. (Revelation 11:3)

This number also has a two fold implication -- a good and an evil sense. In its favorable implication it speaks of covenant (an agreement between two parties), help, and hence salvation (one helping another one). Thus it stands for the second, or redemptive,

Person of the Trinity. In its evil sense it implies division, separation, and strife.

When the earth lay in chaos (Genesis 1:2), its condition was universal. RUIN and DARKNESS formed a blanket over the universe. The SECOND thing recorded in connection with CREATION was the introduction of of a SECOND THING -- LIGHT; and immediately there was a DIFFERENCE and DIVISION, for God Divided the LIGHT from the DARKNESS.

DIVISION is the characteristic of the SECOND DAY. (Genesis 1:6)

> "Let there be a firmament in the midst of the waters, and let it DIVIDE the waters from the waters."

Here we have DIVISION connected with the SECOND day, and affirming its meaning.

This great significance of TWO, in its SPIRITUAL SENSE, is maintained throughout the Word of God.

It is indeed significant, that the SECOND of any number of things always bears upon it the stamp of DIFFERENCE, and generally of enmity. Take the second statement in the Bible. The FIRST said,

> "In the beginning God created the heavens and the earth." (Genesis 1:1)

The SECOND is,

> "And the earth was (or became) without form and void." (Genesis 1:2)

The FIRST speaks of UNITY and order. The SECOND of DIVISION AND DESOLATION.

The SECOND section of the book of Genesis (Genesis 2:4-4:26) contains the account of the fall; the entrance of the SECOND being -- the Enemy -- "the old serpent", the Devil, introducing discord, sin and death. Enmity is seen first in the SECOND division.

> "I will put enmity between thy seed and her seed." (Genesis 3:15)

We see a SECOND, and enemy in the serpent; a SECOND creature in the woman who was deceived and "in the transgression"; a SECOND man in the seed of the woman (prophecy of Christ). The number TWO becomes associated with Incarnation, "The Second Man, the Last Adam".

If we look at the FIVE books of Moses as a whole, we see in the FIRST book, DIVINE SOVEREIGNTY, UNITY, but the SECOND book (Exodus) opens with "the OPPRESSION of the ENEMY". Here, again, is seen "Another", the Deliverer and Redeemer, who says,

"I am come down to deliver." (Exodus 3:8)

The Old Testament is divided into three divisions, called the LAW, the PROPHETS, and the PSALMS. The SECOND division, the Prophets (Joshua, Judges, Ruth, I and II Samuel, I and II Kings, Isaiah, Jeremiah, and Ezekiel) contains the record of Israel's ENMITY to God and of God's CONTROVERSY with Israel. In the FIRST book of this division (Joshua) we have God's SOVEREIGNTY, UNITY, in giving the conquest of the land; while in the SECOND (Judges) we see the REBELLION and ENMITY in the land, leading to departure from God and the oppression of the enemy. Here we have the "saviour" whom God raised up to deliver His people.

In the THIRD division of the Old Testament, called "The Psalms", because it begins with the Book of Psalms, we have in the Hebrew Canon, as the SECOND book, the book of Job. Here, again, we have seen the ENEMY in all his power opposing and oppressing a child of God.

The same significance of the number TWO is seen in the New Testament. Wherever there are TWO Epistles, the SECOND has some special reference to the ENEMY.

In II Corinthians there is a marked emphasis on the POWER of the ENEMY, and the working of Satan.

In II Thessalonians we have a special account of the working of Satan in the revelation of the "man of sin" and "the lawless one".

In II Timothy we see the church in its RUIN, as in the FIRST epistle we see it in its RULE.

In II Peter we have the coming apostasy foretold and described.

In II John we have the "antichrist" mentioned by this name, and are forbidden to receive into our house any who come with his doctrine.

It is also to be found in the vast number of things which are introduced in PAIRS, so that the one may teach concerning the other by way of contrast or difference.

The TWO Sons. (Matthew 21:28, and Luke 15:11, and Galatians 4:22)

The TWO Foundations. (Matthew 7:24-27: "the one which fell NOT, for it was founded on a ROCK"; the other, which "fell, and great was the fall of it."

What may be called the great separation chapter of the Old Testament is found in Genesis nineteen, where we have the separation of Lot and his daughters from those who are destroyed in their sins. The chapter is full of the figure TWO.

In verse 1 we have "TWO angels".

In verse 4 we have "TWO classes of age," "old and young."

In verse 8 we have "TWO daughters."

In verse 11 we have TWO classes of standing, "small and great".

In verse 17 we have TWO places, "the plain", the place of danger, and the mountain", the place of safety.

In verse 24, we have TWO cities destroyed, Sodom and Gomorrah.

In the same verse we have TWO agents of destruction, "brimstone and fire."

In verse 3 and 30 we have TWO dwelling - places; verse 3, "his house", and verse 30 , "a cave."

In verse 37 and 38 we have TWO children, Moab and Ben-Ammi.

The corresponding chapter in the New Testament is in Matthew 7, where we find in verse 3 the man with the mote, and the man with the beam.

In verses 13 and 14 the strait gate and the wide gate; the narrow way and the broad way; the way which leadeth unto life, and the way which leadeth unto destruction.

In verse 17 the good tree and the corrupt tree.

In verse 24-27 the wise man and the foolish man; the house built on the rock, and the house built on the sand; the house which stood, and the house which fell.

It is noticeable how often our Lord makes use of this number in His parables, as for instance in the TWO debtors, the tares and the wheat, the rich man and Lazarus, the Pharisee and the publican, the TWO sons of Matthew 21:28.

All through the Bible too, there are instances of TWO persons, of ONE whom ONE was divided off from the other. Such are Cain and Abel, Ishmael and Isaac, Esau and Jacob, Orpha and Ruth, Vashti and Esther. James and John. Note also that the FIRST always comes under the curse while the SECOND is delivered from the curse. How true this is in our life. All mankind is of the FIRST born -- under Adam -- and we are under the curse. But as we receive Christ as our Saviour, He becomes the cursed for us, "The first begotten of the Father" that we might be delivered from the curse and receive the life of God. In Him we are delivered from the FIRST curse, and the FIRST Death, and the FIRST EARTH that we might receive the SECOND BIRTH, from the SECOND man, Who delivers us from the SECOND death at His SECOND coming.

Always when those who have been ONE body are separated into TWO bodies there is a division of some kind. It may be a peaceful DIVISION, but it is DIVISION none the less. More often than not it is not a peaceful DIVISION.

Genesis 10:25 says,

> "And unto Eber were born TWO sons; the name of one was Peleg; for in his day was the earth DIVIDED; and his brother's name was Joktan."

Why should this seemingly unimportant passage be in the Word of God, f not to teach a truth? Here the number TWO is connected with DIVISION.

There was DIVISION between the first TWO sons who were born in the world, Cain and Abel. (Genesis 4:1-10) The same was true with Abraham's TWO sons, Ishmael and Isaac. (Genesis 21:8-13) The same was true with Isaac's TWO sons, Esau and Jacob. (Genesis 25:27-34 and 27:41-45).

In Genesis 15:10 the word rendered "DIVIDED", in the Hebrew is "BATHAR", which only occurs twice in this verse. It is the chapter where the "word of the Lord" first occurs, and is mentioned twice in verse 1 and 4. Then we are told that Abram "believed" that Word, and "He counted it to him for righteousness," and thus we get a picture of how faith in God's Word links us on to Christ's righteousness and divides us from the unbelievers. Then the animals are "divided", and Abram is given God's great unconditional promise of the Land to his seed, which is now in these last days just about to be fulfilled, when those who have been separated by faith as Abraham was, will then be finally separated in resurrection from the unsaved, and will reign with Christ over a sin-cleansed earth.

In Matthew 6:24 Jesus said,

> "No man can serve TWO masters: for either he will hate the one and love the other, or else he will hold to the one and despise the other."

Here the number TWO is connected with a DIVIDED service and devotion, which Jesus says is impossible.

In James 1:8 it is said,

> "A double minded man is unstable in all his ways."

A double minded person is DIVIDED within himself.

In the days of David and Solomon all the tribes of Israel were united into ONE nation. But after the days of Solomon they were divided into TWO nations. (I Kings 12:1-7) But Ezekiel foretells a time coming when the DIVISION will be abolished and the TWO shall become ONE nation again.

> "Thus saith the Lord God; Behold I will take the children of Israel from among the heathen whither they be gone, and will gather them on every side, and bring them into their own land: and I will make them ONE nation in the land upon the mountains of Israel; and one king shall be king to them all: and they shall be no more TWO kingdoms (nations), neither shall they be DIVIDED into TWO kingdoms any more at all." (Ezekiel 37: 21-23)

Here the number TWO is again connected with DIVISION. Jesus said that He came to bring DIVISION.

> "Suppose ye that I am come to bring peace on the earth? I tell you, Nay; but rather DIVISION; for from henceforth there shall be five in one house DIVIDED, three against two, and two against three." (Luke 12:51-52)

Here is a household DIVIDED into TWO camps with three on one side and two on the other. Oftentimes households are seen DIVIDED because of the grace of God and the truth of the Lord. The truth of God DIVIDES because those blinded by Satan, the god of the world, will not receive the truth. There is nothing wrong with the truth of God or the grace of God, but there is something vitally wrong with those whom Satan deceives, and who will not receive the truth. There would be no division if they would break away from error and receive the truth. There are TWO opposing spirits in the world: the Spirit of Christ, and the spirit of the devil. All people are under the influence of one or the other of these TWO. Consequently, the DIVISION which this brings will continue as long as Satan's influence is among men. Not until he is bound and his influence is stopped for a thousand years can there be peace on earth. The first coming of Jesus did not bring peace, but DIVISION. And we need not expect peace until Christ comes the second time. At His second advent to the earth He will bring peace.

Then the prophecy of the angel's song, "Glory to God in the highest, and on earth PEACE to men of good will," will have its fulfillment. In the meantime, we may expect homes, families, communities and nations to be DIVIDED over the teachings of Christ.

DIVISION and SEPARATION are again shown by number TWO in Luke 17:34-36. Here it is shown THREE times over.

> "I tell you, in that night there shall be TWO men in one bed; the one shall be taken and the other left. TWO women shall be grinding together; the one shall be taken and the other left. TWO men shall be in the field; the one shall be taken and the other left."

Since TWO is the first divisible number in the Bible and in life, it expresses TWO thoughts for us to consider.

1. The thought of REDEMPTION is expressed in the Number TWO.

   In Exodus 8:23 we read, "I will put a "division" between My people and thy people". The word "division" expresses separation, and the Lord separated the Children of Israel from the Egyptians by blood -- the price of redemption.

   TWO angels went to separate Lot from Sodom in Genesis 19. Note how many times through the Bible there is a comparison of righteousness with unrighteousness, light against darkness, sin against salvation, bad against good, and Satan against the Son.

The world is made up of contrasts. We think of many things by way of contrasts - heaven and hell, God and man, saved and lost, salvation and damnation, broad road and narrow road, black and white.

2. TWO also expresses WITNESS-BEARING.

   "Out of the mouth of TWO witnesses every word shall be established". (Matthew 18:16)

TWO angels witnessed to the Resurrection and to the Ascension of our Lord. (Luke 24:4; and Acts 1:10-11)

There were TWO witnesses before the Flood -- Enoch and Noah.

There were TWO witnesses in the Wilderness -- Moses and Aaron.

There were TWO witnesses who bore true witness among the spies -- Caleb and Joshua.

Many others are referred to in the Scriptures. We must always keep in mind that TWO suggest division, or TWO becoming ONE, or redemption and witnessing.

It will illustrate this to state that to bear true witness, we must be separated from the world, and in union with Christ.

There are TWO natures within the believer. If sin is not to have dominion over us, we must realize the truth of separation.

It was the Second Person of the Trinity who was separated from God when He was hanging on the Cross as Sin-bearer, (He cried, "My God, My God, why hast Thou forsaken Me?") and through whom we are separated from the world unto God.

At the second coming of Christ we shall be called out of this world and separated from the lost and be delivered from the Tribulation Period in the fellowship of the Son of God.

This number, TWO is used 808 times in the Bible.

THREE

RESURRECTION, DIVINE COMPLETENESS AND PERFECTION

When a man found in his primitive home the divinest thing that life had to offer -- a father's love, a mother's love and a child's love -- he found God reflected in the interplay of love and kindness and affection in his own household and began to think of the number THREE as a symbol of the Divine. The divinest thing in life was "THREE" and the divine origin of life was THREE. Here in the ultimate world were father's love, mother's love and child's love. Here, too, were the glimpses of the great mysteries which we express in the terms "Father," "Son", and "Holy Spirit". THREE came to carry the thought of the Divine. It means "GOD IS IN IT". It is the number of DIVINE COMPLETENESS AND PERFECTION.

It is the number of the Trinity and is the first of the four perfect numbers (the other three being SEVEN, TEN, AND TWELVE).

The relation of Jehovah to eternity is given in the THREE-FOLD expression "Who is" (present), "Who was" (past), and "Who is to come" (future).

In reference to our Lord many things could be said.

1. THREE times a voice from heaven spake to Him.
2. His great temptation in the wilderness came in a THREE-FOLD way (Luke 4: 3, 6-7, 9-10) being repeated in the three later temptations of His life.
3. He is THE WAY, THE TRUTH, THE LIFE. (John 14:6)
4. He raised THREE from the dead during His earthly ministry. (The widow's son, Jairus's daughter, and Lazarus.)
5. He was crucified at the THIRD HOUR.

6. There was THREE hours of darkness when He was on the Cross.
7. He arose the THIRD day.

THREE is called the "DIVINE NUMBER" because it is mentioned so often in connection with Holy Things. It speaks of the Trinity of God -- Father, Son, and Holy Spirit; and the Trinity of Man -- Body, Soul, and Spirit; and of the Three Great Feast Days, the Passover, Pentecost and Tabernacles; the THREE-FOLD character of the Baptismal Formula, (Matthew 28:19); Jesus THREE prayers in the Garden of Gethsemane; the THREE Denials of Peter and the Lord's THREE-FOLD Question and Charge to Peter.

The number THREE is also associated with the Restoration of Israel. (Hosea 6:1-2) Note the reference to the Resurrection of Jonah and the Resurrection of Christ. (Matthew 12:38-40)

The number THREE is very prominent in the THREE-FOLD ascriptions in the Book of Revelation. Jesus Christ is spoken of as "He which is and was and is to come; as the FAITHFUL WITNESS, the FIRST BEGOTTEN OF THE DEAD, and the PRINCE OF THE KINGS OF THE EARTH." The Four Living Creatures chant, "Holy, Holy, Holy" unto the Almighty, and give Him "Glory, and Honor and Thanks". The Book of Revelation is divided into THREE parts. There are THREE "Woe Trumpets", and THREE Frog-like Spirits issue from the mouth of the Dragon, the mouth of the Beast, and the mouth of the False Prophet. (Revelation 16:13, 14)

THREE plagues are to come upon Babylon, "Death, Mourning, and Famine," (Rev. 18:8) and THREE classes of persons shall bewail her downfall, "Kings, Merchants, and Seamen". These are but a few specimens of the use of the number THREE in Revelation.

The number THREE is also prominent in NATURE. It is the number of MANIFESTATION. The Primary colors of "Solar Light" are Blue, Yellow and Red; and the Sun itself is a Trinity whose manifestations are Light, Heat and Chemical Rays. In nature there are THREE Kingdoms; Animal, Vegetable, and Mineral.

Matter exists in THREE forms; Gaseous, Liquid, and Solid, and the great forces of nature are Gravitation, Light and Electricity. Time is divided into THREE parts: Past, Present and Future.

The history of the Earth between the Fall of man and the Renovation of the Earth by Fire is divided into THREE ages, the Antediluvian, Present, and the Millennial age.

The fullest human life is lived in the THREE planes of body, mind, and spirit and there are THREE sections to the human brain; the Hind Brain, Mid-Brain and the Fore-Brain. These are also called the cerebrum, the upper main part; and a hinder part, the cerebellum and the lowest part of the brain called the medulla oblongata, which is continuous through the foramen magnum with the spinal cord. The brain is invested by THREE membranes. These are an outer, the dura mater, and the pia mater. All of life revolves around the number THREE.

The Gospel of Christ is THREE-FOLD: The death, burial and resurrection of Christ. (I Corinthians 15: 3-4). It saves from the past, sanctifies for the present, and glorifies in the future.

No other number in all the Bible reveals the works of the Creator than the number THREE.

The number THREE suggests COMPLETENESS, and was often used with a glance at that meaning in daily life and daily speech. Only a selection from the great mass of Biblical examples can be given here. It would be well for the reader to study the many examples given throughout the Word of God.

1. THREE is often found of persons and things sacred or secular. Noah's THREE Sons (Genesis 6:10); Job's THREE daughters (Job 1:2; 42:13) and THREE friends (Job 2:11); Abraham's THREE guests (Genesis 18:2); and Sarah's THREE measures of meal (Genesis 18:6); the THREE night watches (Judges 7:19); and God's THREE-FOLD call of Samuel. (I Samuel 3:8)

2. In a very large number of passages THREE is used of periods of time. THREE days; THREE weeks; THREE months and THREE years. (Notice Genesis

40:12, 13, 18; Exodus 2:2; Isaiah 20:3; Jonah 1:17; Matthew 15:32; Acts 9:9; II Cor. 12:8.)

3. The number THREE is also used in a literary way, sometimes appearing only in the structure. Note as examples the THREE-FOLD benediction of Israel (Numbers 6:24); the Thrice Holy of the Seraphim (Isaiah 6:3); the THREE-FOLD refrain of Psalm 42:43 regarded as One Psalm (Psalm 42:5, 11; 43:5); the THREE graces of 1st Corinthians 13; and THREE witnesses. (I John 5:8)

In some of these cases THREE-FOLD repetition is a mode of expressing the superlative, and others remind us of the remarkable association of THREE with Deity.

Last of all, we have the mentioning of the THIRD heaven in II Corinthians 12:1-2, in which Paul had seen a vision and was caught up to THIRD heaven and saw the glory of God. Here is a manifestation of the glory of our Lord.

The number THREE also stands for the RESURRECTION of the body. Jesus said,

> "As Jonas was THREE days and THREE nights in the whale's belly; so shall the Son of man be THREE days and THREE nights in the heart of the earth." (Matthew 12:40)

In John 2:19, He said to the Jews,

> "Destroy this temple, and in THREE days I will raise it up."

In Verse 21 Jesus says He was speaking of the temple of His body. By this Jesus taught that His body would be raised from the dead after being dead for THREE DAYS and THREE NIGHTS. Not only did Jesus teach that He would actually be raised from the dead after THREE days, but He also taught that the THREE days and THREE nights Jonah spent in the whale's belly was a type of His death, burial and resurrection. The book of Matthew, the book of John, and the book of Jonah are in agreement on the number THREE being associated with the RESURRECTION of the body.

There are also THREE recorded cases found in the Old Testament of people being raised from the dead. The first was that of the son of the widow of Zarapath. (1st Kings 17:9-24). This child was raised by the prophet Elijah. Verse 21 reads,

> "He (Elijah) stretched himself upon the child THREE times, and cried unto the Lord, and said, O Lord, my God, I pray thee, let this child's soul come into him again."

Why did Elijah stretch himself on the body of the child just THREE times, if not by inspiration? Was it not recorded for our learning? The next verse reveals that the child lived again.

The second case of a person being raised from the dead in the Old Testament is found in II Kings 4:18-35. This time it was the prophet Elisha who raised from the dead the son of the Shumanite woman. In the 34th verse this is recorded,

> "And he (Elisha) went up, and lay upon the child and put (1) his mouth on the child's mouth, and (2) his eyes upon the child's eyes, and (3) his hands upon the child's hands: and he stretched himself upon the child; and the flesh of the child waxed warm."

Later, the child sneezed SEVEN times, and opened his eyes, (verse 35). Why is it recorded that Elisha put (1) his mouth on the child's mouth, and (2) his eyes upon the child's eyes, and (3) his hands upon the child's hands? Is it not to show that the number THREE is the number that is associated with the bodily RESURRECTION?

The third and last case is found in II Kings 13:21. There they cast the dead body into the sepulchre of Elisha and he revived. Why are there recorded THREE and only THREE cases of people being raised from the dead in the Old Testament? Why not only two, or why not four or five or more? Was not the Spirit of God leading in it? Was it accidental?

Isn't it an amazing thing to note that there were THREE RESURRECTED in the Old Testament and THREE RESURRECTED in the New Testament which is SIX, man's number of weakness. The SEVENTH RESURRECTION (Completeness) is the Lord Jesus Christ Himself, and through HIS RESURRECTION all mankind can rise and live forever. In the six, they were RESURRECTED but died again, but in CHRIST they live forever.

In Exodus 12:37 to 13:20, Israel journeyed THREE days before crossing the Red Sea. This is another of the pictures of RESURRECTION. Moses said to Pharoah,

> "We will go THREE days journey into the wilderness, and sacrifice unto the Lord our God, as He commands us." (Exodus 8:29)

Why did God command them to go THREE days? Why not two, four or more?

To complete this picture compare the time of day in Exodus 14:27 with that of Matthew 28:1-6, and behold the wisdom of God, and marvel, and believe.

> "And Moses stretched forth his hand over the sea, and the sea returned to his strength WHEN THE MORNING APPEARED." (Exodus 14:27)

This was when the day was dawning. Israel had emerged from her watery grave and was standing, typically, on RESURRECTION ground. Matthew 28:1 says,

> "In the end of the sabbath, as it began to DAWN (or as the morning appeared) toward the first day of the week, came Mary Magdalene and the other Mary to see the sepulcher."

The record goes on to say they found the grave empty and were told by the angels that He was risen from the dead. Why was the number THREE associated both with Israel's crossing of the Red Sea and the RESURRECTION of Jesus? Why was Israel found out of her watery grave at the same time of day that the women found the grave of Jesus empty? Will the doubter answer this? Then let him read I Corinthians 5:7,

> "Even Christ our passover is sacrificed for us,"

and see that the passover lamb in Egypt was a picture of Christ whom John called,

> "The Lamb of God which taketh away the sin of the world." (John 1:29)

Then let him tell why it was THREE days after the passover lamb was slain that Israel crossed the RED SEA, and THREE days after Christ was crucified that He arose from the dead. Why are these types, figures and numbers so fitting?

THREE times the waters of Jordan were parted. The first time is the case mentioned Josh. 3. The second time is found in II Kings 2:6-8. The THIRD time is II Kings 2:13-14. Why THREE times and only THREE, if there is no meaning to it?

THREE times the RESURRECTION of the Lord's people is connected with Christ's return. (I Corinthians 15:22-23; Philippians 3:20-21; and I Thessalonians 4:16)

THREE times the Lord said about those who believe on Him, "I will raise him up at the last day." (John 6:40, 44, 54)

Let the doubter tell why all these writers associated the number THREE with the RESURRECTION unless they were inspired of God to do so. The law of averages would be against it occuring so many times for it to be accidental. The writers lived too far apart in point of time to have agreed upon this thing and to palm off a deception on the world.

This number is mentioned 467 times in the Word of God.

FOUR

CREATION - - - - WORLD

When man went outside of his home and looked about him, he had no conception of the modern world as we know it. No Copernicus had ever opened his eyes to the vast significance of the universe. To him the world was a great flat surface with four boundaries, east and west and north and south. There were four winds from the four sides of the earth. In the town he placed himself within the limit of four walls. Thus, when he thought of the world he thought in terms of FOUR. FOUR became the cosmic number. The world in which men lived and worked and died, was conveniently symbolized by FOUR.

A number of further divisibility, FOUR stands for the WEAKNESS found in the world and man. In common parlance we speak of "the FOUR corners of the earth" and "the FOUR points of the compass". Important is the indirect meaning of trial, testing, and experience, derived from the fact that the earth is the scene of man's testing.

FOUR is the number of CREATION and marks GOD's CREATIVE WORKS. It is the signature of the world.

The earth has,

1. FOUR regions; North, South, East, West.
2. FOUR elements; earth, air, fire, water.
3. FOUR seasons; Spring, Summer, Autumn, Winter.
4. FOUR kingdoms; mineral, vegetable, animal, spiritual.
5. FOUR winds; from 4 directions of this earth.
6. FOUR divisions of our day; morning, noon, evening, night.
7. FOUR phases of the moon; 1st Quarter, New Moon, Last Quarter, Full Moon.

The FOURTH day the material creation was finished. (Genesis 1:14-19)

In the first and second chapters of Genesis, in the record of CREATION, the word CREATURE is found FOUR times. (Genesis 1:20; 1:21; 1:24; and 2:19)

FOUR is the number that is associated with CREATION. In Romans 8:19-22 the words CREATURE and CREATION are used FOUR times in succession.

> "For the earnest expectation of the (1) CREATURE waiteth for the manifestation of the sons of God. For the (2) CREATURE was made subject to vanity, not willingly, but by reason of him who hath subjected the same in hope, because the (3) CREATURE itself also shall be delivered from the bondage of corruption into the glorious liberty of the children of God. For we know that the whole (4) CREATION groaneth and travaileth in pain together until now."

In the Revised Version the word is CREATION in all FOUR places. Thus, the word CREATION is used FOUR times in succession in four verses. These are the only times this word is used in this chapter.

In Revelation 5:13 the CREATURES in FOUR different places ascribe FOUR words of praise to the Father and to Christ.

> "And every CREATURE which is in heaven (1), and on the earth (2), and under the earth (3), and such as are in the sea (4) and all that are in them heard I saying, (1) blessing, and (2) honor, and (3) glory, and (4) power, be unto him that sitteth upon the throne, and unto the Lamb forever and ever."

In Revelation 4:6-8 John saw FOUR living CREATURES around the throne of God.

> "And in the midst of the throne, and around about the throne, were FOUR beasts full of eyes before and behind."

The word translated "Beasts" in this connection is rendered, "Living Creatures" in all other translations. Notice that there were FOUR of these living CREATURES.

There are FOUR divisions of the human race. The FOUR living CREATURES represent those redeemed from every (1) kindred, (2) people, (3) tongue, and (4) nation. The study of number TWENTY-FOUR will show this number connected with the priesthood (believer priests). The crowns the elders were wearing (Rev. 4:4) show them to be kings.

The River that flowed out from the "Garden of Eden" was divided into FOUR parts. (Genesis 2:10-14)

Ezekiel had a vision of the Cherubims. They were FOUR in number. Each had FOUR faces and FOUR wings. The first face was that of a Man. The second that of a Lion, the third that of an Ox, and the FOURTH that of an Eagle, all of them "earthly creatures" (Ezekiel 1:5-10). Here the number FOUR is associated with CREATURES FOUR times over.

The FOUR world judgments to come upon the Nations at war are, War, Famine, Pestilence and Earthquakes. (Matthew 24:6, 7 and Revelation 6)

The great "World Powers" as revealed to the Prophet Daniel were four in number; Babylon, Medo-Persia, Greece and Rome. These are all earthly powers. They were ruled over by the natural man, represented by number FOUR. Following these FOUR is the Fifth, which will be controlled by Christ and those whom He saved by His grace. FIVE is the number for "grace". (Daniel 2:31-43, and 7:3-23)

In the parable of the sower, which represents the gospel being preached to mankind, Jesus mentioned FOUR places, the (1) wayside, (2) the stony places, (3) the thorns, and (4) the good ground. (Matthew 13:3-8; Mark 4:3-8; and Luke 8:4-8)

There are FOUR portraits of Christ in the FOUR Gospels. MATTHEW reveals Him as the King. Written primarily for the Jew, He is the Son of David. His royal genealogy is given in chapter 1. In chapters 5-7, in the Sermon on the Mount, we have the manifesto of the King, containing the laws of His kingdom. MARK reveals Jesus as the Suffering Servant. Written to the Romans, there is no genealogy. Why? Men are not interested in the genealogy of a servant. More miracles are found in this gospel than in any other Gospel. Romans cared little for words; far more for deeds. Luke reveals Jesus as the Perfect Man. Written to the Greeks, His genealogy goes back to Adam, the first man, instead of to Abraham. As a perfect Man, He is seen much in prayer and with angels ministering to Him. JOHN reveals him as the Son of God. Written to all who will believe with the purpose of leading men to Christ (John 20:31), everything in this Gospel illustrates and demonstrates His divine relationship. The opening verse carries us back to the "beginning".

The Brazen Altar had FOUR sides and FOUR horns. (Exodus 27:1-2)

The New Jerusalem in the new heavens and the new earth is to be FOUR-SQUARE. (Revelation 21:16)

There are FOUR Divine Acts on the first day of the earth's restoration. (Genesis 1:5) God said, God saw, God divided, and God called. There were FOUR Divine Acts on the second day of earth's restoration. (Genesis 1:8) God said, God made, God divided, God called. There were FOUR Divine Acts on the fourth day. (Genesis 1:19). God said, God made, God set, God saw. There were FOUR Divine Acts on the Seventh Day. (Genesis 2:2) God ended, God rested, God blessed, and God sanctified. These FOUR Divine Acts were performed on FOUR of the seven days and was in relationship to man and his existence upon this earth.

The FOURTH Commandment of the Ten Commandments is "Remember the Sabbath Day to keep it holy." (Exodus 20:8) This day was given by God to man to provide him rest for his body upon this earth.

There were FOUR blessings of the righteous. (Psalm 1:3; 32:6; 33:18) FOUR commands to kings and rulers on this earth. (Psalm 2:10-12) There were FOUR coverings for the tabernacle erected in the Wilderness. (Exodus 26:1)

1. The inner material of fine linen curtains,

housing the holy place and most Hóly Place. (Exodus 26:1-6)
2. The tent or middle covering of goat's hair. (Exodus 26:7-13)
3. The first outside covering of ram's skins dyed red. (Exodus 26:14)
4. The outside covering of badger skins to withstand weather. (Exodus 26:14)

There were FOUR rows of stones in the Breastplate of the High Priest that represented the 12 sons of Jacob. (Exodus 28:17-21)

There were FOUR curtains to the entrance of the Tabernacle in the Wilderness with the Altar of Sacrifice just within, and they speak of the FOUR Gospels that witness to the sacrifice of Christ.

There were FOUR colors of curtains, each having a typical significance.

1. The blue speaks of Christ as the Heavenly ONE, or as the Son of God, as revealed in John.
2. The purple speaks of Christ as the Royal ONE, the King, or Messiah, as in Matthew.
3. The scarlet speaks of Christ as the Perfect Servant, obedient unto death, as in the gospel of Mark.
4. The white linen speaks of Christ as the Perfect Man, the Son of Man, as in Luke.

There were FOUR animals that an offerer could bring as a sacrifice unto the Lord.

1. The bullock typifies Christ as the Servant, obedient unto death. (Philippians 2:5-8)
2. The lamb or sheep typifies Christ as the Lamb of God taking away our sins. (Isaiah 53:7; John 1:29; I Peter 1:19)
3. The goat typifies Christ as being made a curse (Galatians 3:13 and a transgressor. (Isaiah 53:12)
4. The turtledoves typify Christ as the One

who became poor that through His poverty we might be rich. (II Corinthians 8:9; and read also Luke 2:24)

Number FOUR then, is the number of the World and its people. It is found 305 times in the Scriptures.

FIVE
GRACE OR GOD'S GOODNESS

Next, man turned from the study of his home and the world about him to study himself. Perhaps our decimal system arose from the intensive study by a man of his own fingers and toes. That was a crude and cruel age where many were maimed and crippled through disease, accident, or warfare. A perfect, full-rounded man was one who had all his members intact. When a man looked down and saw his five fingers, five toes and realized he had good health of five senses, he came to realize the GOODNESS OF GOD or the GRACE OF GOD.

Number FIVE is the number of GRACE. Here we have a number which is generally understood as signifying a responsibility upon man, the recipient of God's grace.

The first FIVE books of the Bible, called the Pentateuch, the foundation of Scripture, makes known the righteous requirements of God which man is obligated to obey.

It is a significant number in connection with the measurements of the Tabernacle which was certainly a place where God's grace was manifested.

There were FIVE offerings upon the Altar of Sacrifice.

1. THE BURNT OFFERING.
   This was the principal offering in connection with this altar, in which the offering was burnt to ashes, the priest getting only the skin of the animal. (Leviticus 7:8) The whole offering went up to God and thus speaks of His own satisfaction in the death of Christ; the ashes speak of the absolute and thorough judgment of sin; the skin of the animal, as it was with Adam and Eve (read Genesis 3:21), speaks of the provision God has made in the death of Christ to clothe the naked, guilty sinner in the garments of salvation. (Isaiah 61:10 and Luke 15:22)

2. THE PEACE OFFERING.
   This offering typifies Christ making peace by the blood of His cross (Colossians 1:20 and Ephesians 2:14-17). As priests, believers are now to feed upon the affections and strength of Christ, as typified in the breast and shoulders, respectively. (Read Leviticus 3:1-17 and Leviticus 7:11-21)
3. THE SIN OFFERING.
   In this offering we see Christ typified as our substitute, bearing our judgment due to our sins, answering to the righteous requirements of the law. (Read I Peter 2:24; 3:18; Psalm 22 and Isaiah 53. Leviticus 4:1-35 and Leviticus 6:25-30)
4. THE TRESPASS OFFERING.
   This typifies Christ's death as atoning for every effect of sin, rather than for guiltiness, and as the basis upon which God will eventually effect the restitution of a disordered universe. This offering was twofold, covering trespasses both against God and against one's neighbor. (Leviticus 5:1-19 and Leviticus 6:1-7)
5. THE MEAT OFFERING.
   This offering typifies Christ's perfect humanity, already partially referred to; the frankincense speaks of Christ's fragrant life (Matthew 3:17); the oil, Christ's anointing of the Spirit for His public ministry (Matthew 1:18-23); the oven speaks of the last three hours of His suffering on the cross, unseen (Psalm 22, Hebrews 2:18) the pain, His sufferings at the hands of men, seen in Matthew 27:27-31.

FIVE ministries through which God's grace is to be revealed are given in Ephesians 4:11, namely -- Apostles, Prophets, Evangelists, Pastors and Teachers.

David chose FIVE smooth stones when he went against the giant Goliath -- He trusted in God's grace. There were FIVE giants of the nation of the Philistines and

David went down to get a complete victory for the Lord. (I Samuel 17:40)

We have two sets of Commandments of FIVE each. FIVE in relation to God, and FIVE in relation to our fellowman.

The Psalms contain FIVE Books, each closing with the doxology.

1. Book One. Psalms 1-41
   The first book of Genesis: Concerning Man.
2. Book Two. Psalms 42-72
   The Exodus Book: Concerning National Israel.
3. Book Three. Psalms 73-89
   The Leviticus Book: Concerning the Sanctuary.
4. Book Four. Psalms 90-106
   The Numbers Book: Concerning Israel and Gentiles.
5. Book Five. Psalms 107-150
   The Deuteronomy Book: Concerning God and His Word.

This number is not of frequent occurrence. There were FIVE wise and FIVE foolish Virgins. Jesus fed the multitudes (FIVE thousand men), with FIVE loaves. (John 6:1-10)

This number FIVE is mentioned 318 times in the Bible and always reveals the GOODNESS AND GRACE of our Lord toward man.

There are FIVE women mentioned in the genealogy of Christ in Matthew 1. They are Rahab, Ruth, Tamar, Bathsheba, and Mary. All had experienced the grace of God in their lives.

The Apostle John wrote FIVE books that reveal the Grace of God toward the believers. The Gospel of John, I, II, III John and Revelation.

There are FIVE things a Christian is told to remember.

1. What you are saved from. (Ephesians 2:11-12)
2. Those who suffer. (Hebrews 13:3)
3. Those who rule over you. (Hebrews 13:7)

4. Truth. (Jude 17-18; Revelation 3:3)
5. Backslidings, and repent. (Revelation 2:5)

The word GRACE is used FIVE times in succession in ROMANS 11:5-6. The word is not found again in this chapter. Is there not a significance in this?

> "Even so at this present time also there is a remnant according to the election of GRACE. And if by GRACE, then it is no more of works: otherwise GRACE is no more GRACE. But if it be works, then it is no more of GRACE: otherwise work is no more work."

If the reader will check again this Scripture he will see that the words "WORK" and "WORKS" are found FOUR times. Why is GRACE mentioned FIVE times and "work" FOUR times? GRACE is of God. The works mentioned here are of man. The natural man, man of the first creation, the unsaved man is represented by the number FOUR. He depends upon himself and his own works for salvation. So his work is mentioned FOUR times. If man is saved he must be brought out of the place he occupies, represented by FOUR. He must be brought to GRACE, represented by FIVE. This becomes more enlightening with further study.

Peter addresses his epistle to the saints scattered through FIVE countries, and he mentions FIVE doctrines, and these two FIVES are followed by the word GRACE.

> "Peter, an apostle of Jesus Christ, to the strangers scattered throughout PONTUS, GALATIA, CAPPADOCIA, ASIA, and BITHYNIA,
>
> ELECT according to the FOREKNOWLEDGE of God the Father, through SANCTIFICATION of the Spirit, unto OBEDIENCE and sprinkling of the BLOOD of Jesus Christ:

> GRACE unto you, and peace, be multiplied."
> (I Peter 1:1-2)

How did it happen that Peter addressed this book to saints in FIVE countries, then mentioned FIVE doctrines and then immediately follows with the word GRACE? Was this accidental? Are things accidental in the Word of God?

Note Ephesians 4:1-3 which reads,

> "I therefore, the prisoner of the Lord, beseech you that ye walk worthy of the vocation wherewith ye are called, with all LOWLINESS and MEEKNESS, with LONG SUFFERING, FOREBEARING one another in love; ENDEAVOURING to keep the unity of the Spirit in the bond of peace."

Here we see FIVE things revealed that brings peace. If one has peace one must have GRACE. The apostle, in all his salutations always said, "GRACE be unto you, and PEACE." Look at the beginning of every one of Paul's letters. (Romans 1:7; I Corinthians 1:3; II Corinthians 2:2; Galatians 1:3; Ephesians 1:2; Colossians 1:2; I Thessalonians 1:1; II Thessalonians 1:2; Philippians 1:2; I Timothy 1:2; II Timothy 1:2; Titus 1:4; and Philemon 3)

God promised Jacob FIVE things, (1) to give him the land on which he was sleeping, (2) to be with him, (3) to keep him in all places he would go, (4) to bring him back again, (5) not to leave him. (Genesis 28:13-15) Here is GRACE that gives, GRACE that accompanies, GRACE that keeps, GRACE that brings home, and GRACE that never forsakes. What a beautiful picture of the Security of the Believer in Christ.

There were FIVE ingredients in the holy anointing oil. They were (1) myrrh, (2) sweet cinnamon, (3) calamus, (4) cassis and (5) olive oil. (Exodus 30:23-25) This holy anointing oil was a type of the Holy Spirit by whom Jesus was anointed. (Acts 10:38) In Hebrews 10:29 the Holy Spirit is called "The Spirit of GRACE". There are FIVE references to the Holy Spirit as Comforter. (John 14:14; 14:26; 15:26; 16:7 and Acts 9:31) Marvelous is the Word of God.

Isaiah speaks of FIVE names by which Christ was to be called.

> "His name shall be called (1) Wonderful, (2) Counselor, (3) The Mighty God, (4) The everlasting Father, (5) The Prince of Peace." (Isaiah 9:6)

In John 1:17 John said, "GRACE and truth came by Jesus Christ." The FIFTH name by which He would be called was "Prince of Peace". GRACE brings PEACE.

In Romans 8:29-30 there are recorded FIVE things God did for His people. (1) He foreknew them. (2) He predestinated them. (3) He called them. (4) He justified them. (5) He glorified them.

There were FIVE porches at the pool called "Bethesda". (John 5) This is found in the FIFTH chapter of John. To this place people came to be healed of their infirmities. This pictures GRACE. Jesus asked the question to the lame man at the pool, "Wilt thou be made whole?" There are FIVE words in this question.

Jeremiah foretold the future GRACE that would be upon Israel when he said that in Jerusalem there should be (1) The VOICE of joy, and (2) the VOICE of gladness, and (3) the VOICE of the bridegroom, and (4) the VOICE of the bride, and (5) the VOICE of them that say, Praise the Lord of hosts: for the Lord is good; for His mercy endureth forever." (Jeremiah 33:10-11)

You will note the Scripture in Hebrews 9:14,

1. "How much more shall the blood of Christ,
2. Who through the eternal Spirit
3. Offered Himself without spot to God,
4. Purge your conscience from dead works
5. To serve the living and true God."

There are several things in the FIVE divisions of this verse. In Number ONE is the Son, Jesus Christ, who by the GRACE of God tasted death for every man. The Holy Spirit is number TWO, the Spirit of GRACE (Hebrews 10:29), enabling Christ of offer Himself. The Father, who in I Peter 5:10 is called the God of all GRACE, is in number THREE, accepting the offering of His Son. In number FOUR is the lost man, in dead works, needing GRACE. And in number FIVE he is made alive by GRACE and he can now serve God. The first THREE are the TRINITY. FOUR is man in dead works in this world. In FIVE he is saved by GRACE.

Let those who teach that a lost man must be baptized to be saved take a good look at the divisions of this verse. In number FOUR man is found in dead works. Can such works in any way bring about man's salvation, or help to bring it about? NO!

Read the Scripture found in John 3:14-15.

1. As Moses lifted up the serpent in the wilderness,
2. Even so must the Son of Man be lifted up;
3. That whosoever believeth in Him
4. Should not perish, but
5. Have everlasting life.

Again under number FOUR the sinner is perishing, needing salvation by GRACE. In number FIVE God's GRACE offers eternal life to the perishing man.

This same teaching is found in John 3:16,

1. For God so loved the world,
2. That He gave His only begotten Son,
3. That whosoever believeth in Him,
4. Should not perish, but
5. Have everlasting life.

The first THREE are the Trinity. Number ONE shows the God of all GRACE loving; number TWO shows the Son dying, and by the grace of God tasting DEATH for every man; and number THREE shows the Holy Spirit, the Spirit of GRACE, who enables man to believe. In number FOUR man is perishing, and in number FIVE the GRACE of GOD is offering everlasting life.

Noah's name is found the FIFTH time in Genesis 6:8 where it is said,

> "Noah found GRACE in the eyes of the Lord".

Ruth's name is found the FIFTH time in Ruth 2:2,

> "And Ruth the Moabitess said unto Naomi, Let me go to the field, and glean ears of corn after him in whose sight I shall find GRACE."

In the FIFTH place where the name of Boaz is found he tells Ruth to abide fast by his maidens. Then Ruth said to him,

> "Why have I found GRACE in thine eyes, that thou shouldst take knowledge of me, seeing I am a stranger?" (vs 8-10)

The FIFTH time David's name is found is where Saul sent to Jesse, saying,

> "Let David, I pray thee, stand before me; for he hath found FAVOUR in my sight." (I Samuel 16:22)

Favour means GRACE. What proof could one ask for to show that FIVE stands for GRACE from one end of the Bible to the other? FIVE stands for GRACE. Marvelous is the WORD OF GOD!

SIX

WEAKNESS OF MAN – EVILS OF SATAN – MANIFESTATION OF SIN

In the Bible this number usually has an intensely evil significance, standing for the MANIFESTATION OF SIN. To the Jew the number "SIX" had a sinister meaning. As SEVEN was the sacred number, SIX fell short of it and failed. SIX was the charge that met defeat, with success just in its grasp. It had within it the stroke of doom. It had the ability to be great but failed to measure up. It was for the Jew what THIRTEEN is for many today -- an evil number. It is possible that the dread of this number was revealed one night at the table when THIRTEEN men broke bread and one went out to betray the Master. From that room went one to commit the blackest act in history and one to make the supreme sacrifice of history. Thus SIX was an evil number for the Jews. It is important to keep this in mind when we come to the number "666" in Revelation.

Man was created on the SIXTH day. His appointed days of labor was SIX. (Genesis 1:24-31) (Exodus 20: 8-11) The Hebrew slave was to serve SIX years. For SIX years the land was to be sown and to rest during the seventh. The kingdoms of this world are to last SIX Thousand years.

Moses was compelled to wait for SIX days on the Mount before God revealed Himself to him. (Exodus 24:15-18) SIX days the children of Israel compassed the city of Jericho before its walls fell on the seventh. (Joshua 6:1-2-)

In Daniel 3:1-30, Nebuchadnezzar, a type of those who seek to "deify" man, erected a "Golden Image", typical of himself, and commanded the rulers and people to fall down before it and worship it under penalty of death. The dimensions of the "Image" are worthy of note. It was sixty cubits high and SIX cubits broad. It was prophetic of the "Image of the

Beast'', that the False Prophet will command the people to worship in the Tribulation Period after the Christians have been raptured out of this life. (Revelation 13:3-18)

This number SIX is also connected with SATAN'S INFLUENCE OVER MAN. The SIXTH character in the Bible is the Serpent, which represents satan. The first SIX in the Word of God are: The TRINITY; (3), Adam, (4), Eve, (5), and the serpent (6). (Genesis 1:1-2; John 1:1-2; Genesis 2:21-24; Genesis 3:1-15)

In Revelation 12:9 and 20:2 the serpent is called the Devil, and Satan. Thus the SIXTH character in the Bible is Satan.

The SIXTH time the name of Job occurs in the book of Job it was Satan who used his name, slandering Job by asking a question with SIX words.

> ''Then Satan answered the Lord and said, DOTH -- JOB -- SERVE -- GOD -- FOR -- NOUGHT?'' (Job 1:9)

In Matthew's account of the temptation of Jesus by the Devil, the word ''Devil'' occurs four times, the word ''tempter'' one time and the word ''Satan'' once, making SIX in all. (Matthew 4:1-11). In Luke's account of the same, the word ''Devil'' occurs five times, and ''Satan'' once, making SIX in all. (Luke 4:1-13)

When Jesus was accused of casting out devils by Beezlebub, the prince of the devils, He asked, ''HOW -- CAN -- SATAN -- CAST -- OUT -- SATAN? (Mark 3:22-23) Here are SIX words connected with Satan.

The chief priests rejected Jesus as their king by replying to Pilate with SIX words, ''WE -- HAVE -- NO -- KING -- BUT -- CAESAR.'' (John 19:15)

The SIXTH time the expression ''Thousand Years'' occurs in Revelation 20:1-8 it is connected with Satan.

> ''And when the thousand years are expired, Satan shall be loosed out of his prison, and shall go out to deceive the nations which are in the four quarters of the earth.'' (Verses 7-8)

The work of Satan is seen in connection with the SIXTH vial in Revelation 16:12-14. The work of Satan is also seen in connection with the sounding of the SIXTH trumpet in Revelation 9:13-20.

In the list of the words of the flesh as given by Paul in Galatians 5:19-21 the SIXTH one is "Witchcraft".

Goliath, who challenged the armies of Israel, had SIX pieces of equipment in his armor. They were: (1) an helmet of brass upon his head; (2) a coat of mail made of brass; (3) greaves of brass upon his legs; (4) a target of brass between his shoulders; (5) a spear; and (6) a shield. (I Samuel 17:4-8) When David went out to meet him he took FIVE smooth stones. Here is GRACE overcoming Satan and man. (I Samuel 17:40-45)

The children of Israel in the wilderness lusted after SIX things.

> "And the mixt multitude that was among them fell a lusting: and the children of Israel also wept again, and said, Who shall give us flesh to eat? We remember (1) the fish, which we did eat in Egypt freely, (2) the cucumbers, and (3) the melons, and (4) the leeks, and (5) the onions, and (6) the garlic." (Numbers 11:4-5)

This number SIX might not seem to apply to the Devil in I Timothy 3:16. But this passage, too, is in keeping with all the other passages on number SIX. Here is one of the great truths of the Word of God. The key to it is found in the word "manifest". John said,

> "For his purpose was the Son of God MANIFESTED, that He might destroy the works of the Devil." (I John 3:8)

> "And without controversy great is the mystery of godliness." (I Timothy 3:16)
> 1. God was manifest in he flesh,
> 2. Justified in the Spirit,
> 3. Seen of angels,

4. Preached unto the Gentiles,
5. Believed on in the world,
6. Received up into glory."

How this passage glows with truth when seen in the light of numbers. The first of the SIX statements is "MANIFEST in the flesh." ONE stands for Unity. Jesus declares His ONENESS with the Father, and it stirred up the opposition of the Devil's crowd. In John 10:30 He said, "I and my Father are ONE." "Then the Jews took up stones to stone Him" (verse 31), displaying the opposition of the Devil to His testimony that He and the Father are ONE. "For this purpose was the Son of God MANIFESTED, that He might destroy the works of the devil." (I John 3:8)

The SECOND statement is "Justified in the Spirit". When John baptized Jesus the Holy Spirit came upon Him and pointed out Jesus as the Son of God. John said,

> "I knew Him not: but He that sent me to baptize with water, the same said unto me, Upon whom thou shalt see the Spirit descending and remaining on Him, the same is He that baptizeth with the Holy Ghost. And I saw and bare record that this is the Son of God." (John 1:34-35)

The descent of the Holy Spirit in visible form, like unto a dove, singled out Jesus and SEPARATED Him from all the others as the one who is God's Son. This is the first time in the history of the world that the Holy Spirit found a lodging place upon the earth. Before this, the Holy Spirit came and returned, but now He comes to remain upon Christ. Thus, the Spirit justified His claim as being the Son of God. TWO is the number for Division and Separation. After this He overcame the devil in the wilderness. (Matthew 4: 1-11) Number TWO is also UNION. Here we see the UNION of the Son and the Holy Spirit. TWO as ONE.

The THIRD statement is "Seen of Angels". THREE is the number of God. The Divine Number. It is always associated with the God-head, in creation and resurrection. On the morning of our Lord's resurrection the angels of God were at the tomb and testified to His resurrection. (Luke 24:4-5; Matthew 28:1-7; Luke 24:22-23)

The FOURTH statement is "Preached unto the Gentiles". After His resurrection He was preached unto the Gentiles. FOUR is the number for the world and the lost man. For years controversy has been going on between certain groups as to whether or not the unsaved man is a subject of gospel address. One group contends that the gospel is to be preached to only the unregenerated person. The other group contends that man will be saved whether he wants to or not and the gospel should be preached only to the elect. FOUR is the number that represents man in his natural state in this world, one in the first creation. Here, in the FOURTH statement, it is said that Christ was preached unto the Gentiles, proving the necessity of taking the gospel to lost men everywhere.

The FIFTH statement is "Believed on in the world". FIVE is the number of GRACE. When one believes on Christ, that very moment he is saved by GRACE. Read Ephesians 2:8. Nothing is said about baptism in this statement. "By whom also we have access by faith into this GRACE wherein we stand." (Romans 5:2)

The SIXTH statement is "Received up into Glory". This is the ascension. In Ephesians 1:20-21 Paul tells about Christ being raised from the dead and set at God's right hand in heavenly places, far above all principality, and power, and might, and dominion. Paul makes it plain that the principality and power here is the Devil and his influence.

> "Put on the whole armor of God that ye may be able to stand against the wiles of the Devil. For we wrestle not against flesh and blood, but against principalities, and powers." etc. (Ephesians 6:11-12)

When Christ was received up into glory far above all principality and power, that Manifested His triumph over the Devil. This enables the reader to understand the SIX divisions of I Timothy 3:16 and their connection with the Devil.

The SIXTH time the word "Darkness" occurs in I John 2:11 is:

> "He that hateth his brother is in darkness and walketh in darkness."

When this is read in the light of I John 3;11-12 the work of the Devil becomes more apparent.

> "This is the message that we have heard from the beginning, that we should love one another. Not as Cain, who was of that Wicked ONE, and slew his brother."

The FOURTH time the word "Darkness" is found is "He that saith he is in the light, and hateth his brother, is in darkness even until now." (Verse 9) FOUR is the number of the world, the flesh. So the one who hates his brother is in the flesh, worldly, and of the Devil, and is still in darkness, and has never been in the light, though claiming to be so.

Thus, we find that SIX shows the WEAKNESS OF MAN and the EVILS OF SATAN. This is true all through the Word of the Lord.

SIX is mentioned 199 times in the Bible.

SEVEN

COMPLETENESS - SPIRITUAL - PERFECTION

When man began to analyze and combine numbers, he developed other interesting symbols. He took the perfect world number FOUR and added to it the perfect divine number, THREE, and got SEVEN, the most sacred number to the Hebrews. It was earth crowned with heaven -- the four-square earth plus the divine COMPLETENESS OF GOD. So we have SEVEN expressing COMPLETENESS through union of earth with heaven. This number is used more than all other numbers in the Word of God, save the number ONE.

In the Book of Revelation the number SEVEN is used throughout. There are SEVEN churches, SEVEN Spirits, SEVEN stars, SEVEN seals, SEVEN trumpets, SEVEN vials, SEVEN personages, SEVEN dooms, and SEVEN new things. SEVEN symbolizes Spiritual Perfection. All of life revolves around this number. SEVEN is used over 700 times in the Bible. It is used 54 times in the Book of Revelation.

The whole Word of God is founded upon the number SEVEN. It stands for the SEVENTH day of the Creation Week, and speaks of the Millennial Rest day. It denotes COMPLETENESS or PERFECTION.

In Leviticus 23:15-16, the number SEVEN and the Sabbath, which was the SEVENTH day, is connected with the word COMPLETE. The word COMPLETE follows after the words "SEVEN SABBATH" (Seventh day). The day following the SEVENTH sabbath there was something NEW that took place.

The word FINISHED is also connected with the number SEVEN. In Revelation 10:7 we read,

> "In the days of the voice of the SEVENTH angel, when he shall begin to sound, the mystery of God should be FINISHED."

"It is DONE" is another expression found in connection with the number SEVEN.

> "And the SEVENTH angel poured out his vial into the air; and there came a great voice out of the temple of heaven, from the throne saying, It is DONE." (Revelation 16:17)

The word CREATED is used SEVEN times in connection with God's creative work. (Genesis 1:1; Genesis 1:21; 1:27 (three times); 2:3; and 2:4). God created all things in the beginning and then took six days of restoring His creation and then rested on the SEVENTH day. (Genesis 2:1-3). He appointed SEVEN days for the week, and most, if not all advanced nations reckon time in that way: SEVEN days to the week. Few ever stop to think of why there are SEVEN days in a week. Do atheists and infidels give God and the Bible credit for it?

There are SEVEN notes in the musical scale. All other pitches are only variations of these. When the musician uses the eighth note he goes back to "do" and starts over. Man named the notes but God fixed the sounds, even as God fixed the days of the week, and man named them.

Noah took the clean beasts into the ark by SEVENS. (Genesis 7:2) SEVEN days after Noah went into the ark the flood came. (Genesis 7:9-10) Peter tells about the longsuffering of God waiting in the days of Noah. (I Peter 3:21) Those SEVEN days COMPLETED God's time of waiting.

Before Aaron and his sons entered their priestly work they were consecrated SEVEN days. (Leviticus 8:31-36) Here is a picture of a life COMPLETELY or WHOLLY consecrated or dedicated to the Lord for service.

On the day of atonement the high priest sprinkled the blood upon the mercy seat and before the mercy seat SEVEN times. (Leviticus 16:14) This is a picture of the COMPLETENESS of the redemptive work of Christ.

> "By his own blood he entered in once into the holy place, having obtained eternal redemption for us." (Hebrews 9:12)

When Christ offered Himself that FINISHED the sacrificial offerings. They were ended. No longer must we place sacrifices upon an altar.

When Israel took the city of Jericho God told them to march around the city SEVEN times. Thus, on the SEVENTH day, when they marched around the city SEVEN times, they COMPLETED their marching. (Joshua 6:1-16)

There were SEVEN FEAST days of our Lord. (Passover, Unleavened, First-fruits, Pentecost, Atonement, Trumpets, and Tabernacle). (Leviticus 23:1-44)

There were SEVEN branches on the CANDLESTICK in the Holy Place in the Tabernacle and this pictures the COMPLETE Light of God for the souls of man.

Solomon was SEVEN years in building the Temple and and kept the Feast for SEVEN days. Job had SEVEN sons. When his friends came to visit him they sat SEVEN days and SEVEN nights in silence, and afterward they were required to offer a Burnt Offering of SEVEN bullocks and SEVEN rams. Naaman washed SEVEN times in the Jordan. The Saviour spoke SEVEN words from the Cross. SEVEN men of honest report were chosen to administer the alms of the church in Acts 6:1-7. There were SEVEN years of plenty, and SEVEN years of famine in Egypt during the days of Joseph.

SEVEN times in the Book of Revelation blessings of the Lord are promised to His people. These are called the "BEATITUDES" of Revelation. These are found in Chapters 1:3; 14:13; 16:15; 19:9; 20:6; 22:7, 14.

There are SEVEN Dispensations -- Innocence, Conscience, Government, Patriarchal, Law, Grace, and Millenniah Age.

SEVEN times the Book of Life is mentioned in the Bible. The Book of Revelation is a Book of SEVENS. We have SEVEN churches, SEVEN seals, SEVEN Trumpets, SEVEN Personages, SEVEN vails, SEVEN dooms, SEVEN new things. SEVEN SEVENS make up this Book. It is the COMPLETENESS of all things.

Jesus said to "forgive SEVENTY times SEVEN." In other words, He is saying, "Keep on forgiving until you are complete." Even the duration of Israel's great punishments was based upon this law of SEVENS. Their captivity in Babylon was for SEVENTY years, ten periods of SEVENS. (Jeremiah 25:11-12; Daniel 9:2)

Life operates in a cycle of SEVENS. Changes take place in the body every SEVEN years. There are SEVEN bones in the neck, SEVEN bones in the face, SEVEN bones in the ankle and SEVEN holes in the head. Most births are the multiple of SEVENS. The hen sits three weeks (21 days); the pigeon two weeks, (14 days); after having laid eggs for two weeks. Of 129 species of Mammalia the majority have a period from conception to birth of an exact number of weeks, a multiple of SEVEN.

Human physiology is constructed on a law of SEVENS. Children are born to mothers in a certain number of weeks usually 280 days, a multiple of SEVENS. Fevers and intermittent attacks of gout, ague, and similar complaints have a period of operation of SEVEN, FOURTEEN, or TWENTY-ONE days known as critical days.

All departments of nature are marked all over with mathematics. In this realm practically everything is in SEVENS. Notice next time the shape of frost when it crystallizes on the window. Notice the small snow flakes. It is wonderful how God formed everything in a pattern of SEVENS.

It would be well for the reader to always remember that SEVEN means COMPLETENESS or SPIRITUAL PERFECTION. When ever you come to a SEVEN in the Word of God read the meaning and the message is plain.

SEVEN is found 735 times in the Bible. SEVENFOLD is mentioned 6 times and SEVENTH is found 119 times.

The SEVENTH time Noah's name occurs is where it is said,

> "Noah was a just man, and PERFECT in his generations." (Genesis 6:9)

In the second chapter of Daniel, Nebuchadnezzar had a dream of a great image whose head was of gold, arms and breasts of silver, his belly and thighs of brass, legs of iron, and feet of iron and clay. (Daniel 2:31-33) Daniel told him that he was the head of gold. (vs 37-38) In the next chapter Nebuchadnezzar made an image of gold to be worshipped. At that time he called together SEVEN kinds of officials to the dedication of the image: (1) princes, (2) governors, (3) captains, (4) judges, (5) treasurers, (6) counselors, (7) sheriffs. The penalty for refusing to worship that image was death by being cast into the burning fiery furnace. THREE Hebrew children refused to worship the image and were cast into that furnace of death, and brought forth alive, thus picturing the resurrection, signified by the number THREE. That furnace was heated SEVEN times hotter than it had ever been. Here is a COMPLET DELIVERANCE FOR GOD'S PEOPLE from the power of death. (Daniel 3:1-27) When those THREE Hebrew children came out of that furnace of fire, or death, there was not a trace of fire upon their bodies or their clothing. When Christ brings His people out of death there will not be a trace of death left on them. They will be COMPLETELY delivered from its power.

EIGHT

NEW BIRTH, NEW CREATION OR NEW BEGINNING

The number EIGHT always means a NEW BEGINNING or a NEW ORDER OF THINGS. As the series is complete in SEVEN, EIGHT signifies a new beginning, and thus stands for the NEW in contrast to the old. Observe the EIGHTH day which is really the first day of a new week, or the EIGHTH note of the musical scale which is the same as the first. EIGHT is the number for the NEW BIRTH or the NEW CREATION. FOUR is the number for the first creation and EIGHT is the number for a NEW CREATION.

There were EIGHT persons carried over from beyond the flood in the ark. (I Peter 3:20). With those EIGHT the world was populated ANEW. This is a figure of a new beginning or a new order. This number also carries the thought of RESURRECTION. For EIGHT souls were saved or resurrected in the ark. EIGHT writers in the New Testament speak of the Life, Death, and the Resurrection of our Lord.

On the EIGHTH day our Lord rose from the dead. EIGHT persons were raised from the dead in the Scriptures apart from our Lord Jesus Christ.

In the circumcision of male children of Israel there is a type of the NEW BIRTH.

> "He is not a Jew, which is one outwardly; neither is that circumcision, which is outward in the flesh: but he is a Jew, which is one inwardly; and circumcision is that of the heart, in the spirit, and not in the letter." (Romans 2:28-29)

The male child was circumcised on the EIGHTH day.

> "And he that is EIGHT days old shall be circumcised among you, every man child in your generations." (Genesis 17:12)

David was the EIGHTH son of Jesse. (I Samuel 16:

1-12) Aaron and his sons were consecrated SEVEN days and then began their ministry on the EIGHTH day. (Leviticus 8:31-36)

> "And it came to pass on the EIGHTH day, day, that Moses called Aaron and his sons, and the elders of Israel: and he said unto Aaron, Take thee a young calf for a sin offering, and a ram for a burnt offering . . . and offer them before the Lord." (Leviticus 9:1-2) Then Aaron and his sons entered their priestly work on the EIGHTH day.

The Feast of Tabernacles lasted for SEVEN days, but on the EIGHTH day a "Holy Convocation" was to be held. (Leviticus 23:36) The Feast of Tabernacles was the last of the three great Festivals, and came at the close of the harvest, and during it the children of Israel dwelt in booths. It is typical of God's eternal rest.

When God would have Israel to build the Tabernacle He said to Moses,

> "Let them make me a sanctuary: that I may dwell among them." (Exodus 25:8)

The children of Israel encamped around this Tabernacle in EIGHT groups; FOUR in the outward arrangement, and FOUR in an inward arrangement. The camp of Judah, containing three tribes, was on the east in the outward circle, afar off. (Numbers 2:3-7) On the south was the camp of Reuben, with three tribes. (Numbers 2:10-14) On the west was the camp of Ephraim, with three tribes. (Numbers 2:18-22) On the north was the camp of Dan, with three tribes. (Numbers 2:25-29) In the inward circle there were FOUR divisions of the Levites. They were placed next to the Tabernacle to save the outward FOUR groups from the wrath of God.

> "And the children of Israel shall pitch their tents, every man by his own camp, and every man by his own standard, throughout their hosts. But the Levites shall pitch round

about the Tabernacle of testimony, that there be no WRATH upon the congregation of the children of Israel''. (Numbers 1:52-55)

Notice that the Tabernacle is called the Tabernacle of Testimony.

The third chapter of Numbers gives the FOUR divisions of the Levites who were in the inward circle. The Gershonites were on the west. (vs. 23) The Kohathites were on the south. (vs. 27) The Merarites were on the north. (vs. 33-35) Moses and Aaron and Aaron's sons were on the east before the gate. (vs 38)

In the very center of the camp was the Tabernacle, God's dwelling place, in the midst of the EIGHT groups. This is a picture of Christ dwelling in the heart of the one who has been BORN again. These EIGHT groups were arranged around the Tabernacle to draw a picture of God dwelling in and among those who are BORN AGAIN. EIGHT is the number for the NEW BIRTH or NEW BEGINNING.

In our Lord's conversation with Nicodemus about the NEW BIRTH the word ''BORN'' occurs EIGHT times. (John 3:1-8) In His conversation with the Samaritan woman about the ''living water'' the word ''WATER'' occurs EIGHT times. In this connection the word ''WELL'' is found FIVE times. (John 4:4-15) EIGHT is the number for the NEW BIRTH, and FIVE is the number for GRACE. This shows that the NEW BIRTH is produced by the LIVING WATER, not by water baptism. Jesus never mentioned baptism to either Nicodemus or the Samaritan woman.

The EIGHTH time Noah's name occurs it is said, ''And Noah walked with God.'' (Genesis 6:9) This was a new beginning to him. How wonderful is the Lord and His Word. This number is mentioned 80 times in the Bible.

NINE

FRUIT OF SPIRIT - DIVINE COMPLETENESS FROM THE LORD

The number NINE speaks of FINALITY or DIVINE COMPLETENESS FROM THE LORD. Significantly it is three times three. NINE is the number for the FRUIT OF THE SPIRIT. NINE comes after EIGHT, which represents the NEW BIRTH. After having a good tree, the next thing to expect is good fruit from that tree. "Make the tree good, and his fruit good." (Matthew 12:33) The good fruit follows as the result of the tree being made good. The tree (man) is made good in the NEW BIRTH. As NINE follows EIGHT, so the good fruit, the FRUIT OF THE SPIRIT, follows as the result of the NEW BIRTH.

Paul mentions NINE "fruit of the Spirit" in Galatians 5:22-23,

> "But the Fruit of the Spirit is (1) love, (2) joy, (3) peace, (4) longsuffering, (5) gentleness, (6) goodness, (7) faith, (8) meekness, (9) temperance: against such there is no law."

Here NINE things are mentioned as the FRUIT OF THE SPIRIT. How wonderful and fitting is the Word of God in every way! With what infinite wisdom did He order and arrange His work! Who can gainsay these things? What infidel or atheist can meet or refute such wisdom?

In I Corinthians 12:8-10 Paul mentions NINE gifts of the Spirit.

> "For to one is given by the Spirit (1) the word of Wisdom; to another (2) the word of knowledge by the same Spirit; to another (3) faith by the same Spirit; to another (4) the gifts of healing by the same Spirit;

> to another (5) the working of miracles; to another (6) prophecy; to another (7) discerning of spirits; to another (8) divers kinds of tongues; to another (9) the interpretation of tongues."

In Matthew 5:3-12 there are NINE Beatitudes which our Lord spoke in His sermon on the Mount. These reveals what God wants us to be and the Divine Completeness of the Kingdom of Heaven.

It is also important to remind our readers that it was the NINTH hour that our Lord died on the Cross. (Matthew 27:46-50) Here is recorded the finishing of Sin, Satan, and the COMPLETENESS OF GOD'S PLAN OF SALVATION for man's sin.

This number is used 49 times in the Bible.

The breadth of the court of the Tabernacle was fifty cubits (Exodus 27:12), and there were ten pillars on the west side.

> "And for the breadth of the court on the west side shall be hangings of fifty cubits: their pillars ten and their sockets ten."

Between those ten pillars there were NINE spaces. The Holy Spirit came fifty days after Christ arose from the dead, showing FIFTY to stand for the Holy Spirit and His work. The NINE spaces in the hanging that was FIFTY cubits connects the NINE with the work of the SPIRIT, and shows the FRUIT OF THE SPIRIT.

In the law concerning the sabbath year and what follows, there is both a picture of the NEW BIRTH and also the FRUIT OF THE SPIRIT. God said to Israel,

> "When ye come into the land which I give you, then shall the land keep a sabbath unto the Lord. Six years thou shalt sow thy field, and six years thou shalt prune thy vineyard, and gather in the fruit thereof: but in the seventh year shall be a rest unto the land, a sabbath for the Lord: thou

> shalt neither sow thy field, nor prune thy vineyard." (Leviticus 25:2-4)

In the same chapter they are told what they should eat in the SEVENTH, and EIGHT, and the NINTH years.

> "And if ye say, What shall we eat the seventh year? Behold we shall not sow, nor gather in our increase: then I will command my blessing upon you the sixth year, and it shall bring forth fruit for three years. And ye shall sow the EIGHTH year, and eat of the fruit until the NINTH year; until her fruit come in ye shall eat of the old store." (vs 20-22)

This shows that in the SIXTH, SEVENTH and EIGHTH years they ate of what was planted the sixth year. In the EIGHT year they sowed again, and in the NINTH year they began to eat of the FRUIT of what was sown the EIGHT year. This is a picture of the FRUIT OF THE SPIRIT, represented by number NINE, which follows the NEW BIRTH, represented by number EIGHT.

How could any one deny the meaning that is found in the numbers of the Bible?

TEN

TESTIMONY - LAW AND RESPONSIBILITY

As the basis of the decimal system, TEN has been a significant number in all historical ages. This is a number of TESTIMONY. Considered as twice FIVE, TEN stands for responsibility, intensified, signifying the measure of responsibility and its judgment or reward. This is a number used under the LAW. Man was responsible under the LAW to keep the commandments and bear a testimony for God. This number is made up of the sum of the world number FOUR and SIX the number of man. It is looked upon as a complete number as THREE, SEVEN and TWELVE. Man has TEN digits on his hands and feet. There were TEN Patriarchs before the flood. (Genesis 5) God gave the TEN commandments to man for him to bear Testimony before God and man. There were TEN Plagues upon Egypt and Pharoah during the days of Moses. (Exodus 7:12) Abraham prayed for TEN righteous people within the wicked city of Sodom. He wanted a TESTIMONY FOR GOD. In the parable of the TEN virgins, it gives the legal number necessary for a Jewish function or wedding.

We have the TEN servants to whom were entrusted TEN pounds and one was rewarded by being given authority over TEN cities. (Luke 19:13,17)

TEN powers become powerless against the LOVE OF GOD. (Romans 8:38)

> "For I am persuaded, that (1) neither death, (2) nor life, (3) nor angels, (4) nor principalities, (5) nor powers, (6) nor things present, (7) nor things to come, (8) nor height, (9) nor depth, (10) nor any other creature, shall be able to separate us from the love of God, which is in Christ Jesus our Lord." (Romans 8:38-39)

TEN vices which exclude from the Kingdom of God are listed in I Corinthians 6:9-10.

> "Know ye not that the unrighteous shall not inherit the kingdom of God? Be not deceived: neither fornicators, (2) nor idolaters, (3) nor adulterers, (4) nor effeminate, (5) nor abusers of themselves with mankind, (6) nor thieves, (7) nor covetous, (8) nor drunkards, (9) nor revilers, (10) nor extortioners, shall inherit the kingdom of God."

The Tithe is a TENTH of our earnings belonging to the Lord. We are commanded to "Bring ye all the tithes into the storehouse, etc." (Malachi 3:10) The Tithe is our TESTIMONY of faith unto the Lord.

TEN times we read in Genesis One, "GOD SAID" Here we have the TESTIMONY from the Lord concerning His Creation, and Power.

TEN Psalms begin with Hallelujah. (Psalms 106, 111, 112, 113, 135, 146, 147, 148, 149, 150) Here the Psalmist is giving his TESTIMONY of Praise unto the Lord.

A TEN kingdom confederation is the last phase of human sovereignty upon this earth. This will take place during the TRIBULATION PERIOD. (Daniel 2 and 7; and Revelation 13 and 17)

In Galatians 4:4-5 it is said,

> "When the fullness of time was come, God sent forth His Son, made of woman, made under the Law, to redeem them that were under the Law, that we might receive the adoption of sons."

This proves that the Passover lamb was typical of Christ. In I Corinthians 5:7, the Bible says, "Christ is our Passover Lamb." In Exodus 12:3, the passover lamb is connected with the number TEN.

> "Speak ye unto all the congregation of the children of Israel, saying, in the TENTH day of this month they shall take to them every man a lamb."

Why was the lamb taken on the TENTH day of the month?

Was it not to typify Christ who was to be made under the Law, so He could redeem those who are under the Law so that we could bear testimony for Him?

The lamb was taken up on the TENTH day. (Exodus 12:6) This was the identical day that Christ made His triumphal entry into Jerusalem before the Cross. Here is RESPONSIBILITY and TESTIMONY.

When the meaning of this number is applied to the Tabernacle this becomes more apparent. The boards of the Tabernacle were TEN cubits long. There were TEN linen curtains over the top, and TEN pillars on the west side of the court. (Exodus 26:1, 26:16; 27:12) This shows Christ, made under the Law, fulfilling the Law, and redeeming His people from the Law, so that they can testify of His offering for their sins.

Paul said, "The Law worketh WRATH." (Romans 4:15) In Galatians 5:19-21 Paul lists the works of the flesh. The TENTH one in the list is WRATH.

The number TEN is used 242 times in the Bible and the word "TENTH" is mentioned 79 times.

ELEVEN
JUDGMENT AND DISORDER

The number ELEVEN is associated usually with DISORDER AND JUDGMENT all through the Bible. ELEVEN is one more than TEN. Number TEN represents the LAW and RESPONSIBILITY. A broken Law and Responsibility always brings JUDGMENT and DISORDER. This number ELEVEN is used 24 times in the Word of God.

There were ELEVEN JUDGMENTS upon the Egyptians. Those are as follows:

1. The plague of Blood. (Exodus 7:19-21)
2. The plague of Frogs. (Exodus 8:1-7)
3. The plague of Lice. (Exodus 8:16-17)
4. The plague of Flies. (Exodus 8:21-24)
5. The plague of Murrain. (Exodus 9:1-7)
6. The plague of Boils. (Exodus 9:8-11)
7. The plague of Hail. (Exodus 9:22-25)
8. The plague of Locusts. (Exodus 10:12-15)
9. The plague of Darkness (Exodus 10:21-23)
10. The plague of First-born. (Exodus 12:29-30)
11. The overthrow at the Red Sea. (Exodus 14:24-28)

Israel was delivered from the JUDGMENT that fell upon the first-born of the Egyptians by the blood of the passover lamb. That was the TENTH JUDGMENT. TEN represents the LAW, the TESTIMONY. This pictures deliverance from the condemnation of the LAW when faith is exercised in the blood of Christ. This represents the salvation of the soul. But after the soul is saved there is still the question of security. This is pictured by the protection of the Israelites from the Egyptian army by the pillar of cloud and fire which stood between them.

"The angel of God which went before the

> camp of Israel removed and went behind them; and the pillar of cloud went before their face; and stood behind them: and it came between the camp of the Egyptians and the camp of Israel: and it was a cloud of darkness to them: but it gave light by night to these: so that the one came not near the other all the night." (Exodus 14:19-20)

This shows a picture of the eternal security of God's children. Among the seventeen things that Paul said should not separate God's children from His love he mentions principalities and powers.

In addition to the salvation of the soul and its eternal security there is the salvation of the body, which will take place at the resurrection of the body. This was pictured by Israel's passage of the Red Sea. It was here that the ELEVENTH judgment came upon the Egyptians. Baptism in water pictures the same thing.

Noah pronounced JUDGMENT upon Canaan, a son of Ham, because Ham saw his father's nakedness when he was uncovered in his tent. Noah said,

> "Cursed be Canaan; a servant of servants he be unto his brethern." (Genesis 9:20-25)

In Genesis 10:15-18 it says that Canaan had ELEVEN sons.

> "And Canaan begat (1) Sidon his first-born, and (2) Heth, and the (3) Jubusite, and the (4) Amorite, and the (5) Girgastie, and the (6) Hivite, and the (7) Arkite, and the (8) Sinite, and the (9) Arvadite, and the (10) Zemarite, and the (11) Hamathite; and afterward were the families of the Canaanites spread abroad."

This should answer those who claim that God had nothing to do with the curse pronounced upon Canaan.

In Jeremiah 52:1 it is said that Zedekiah reigned ELEVEN years in Jerusalem. He was a wicked king.

"And he did that which was evil in the sight of the Lord." (verse 2) In Verse 5 Jerusalem was besieged by the king of Babylon until the ELEVENTH year of Zedekiah's reign. Then Zedekiah was captured and taken to Babylon and JUDGMENT was given upon him. (verse 7-9) Again, ELEVEN is shown connected with JUDGMENT.

At Kadesh, sometimes called Kadesh-Barnea, the children of Israel brought JUDGMENT upon themselves by refusing to go up and possess the promised land after the twelve spies had returned. They were condemned to wander in the wilderness until forty years were over, and until all who were 20 years old and upward when they had been numbered had died, with the exception of Joshua and Caleb. (Numbers 13:25 to 14:31). Now read Deuteronomy 1:2:

> "There are ELEVEN days journey from Horeb by the way of Mount Seir unto Kadesh-Barnea."

Those ELEVEN day's journey brought them to the place where JUDGMENT was passed upon them. The country of Sinai is often called Horeb. (Deut. 5:2-27 and Exodus 19:1 to 20:19). At Horeb, or Sinai is where the LAW, represented by the number TEN, was given. ELEVEN is one more than TEN. How fitting it is that the number ELEVEN should be found right after Horeb, and that those ELEVEN days brought Israel to a place of JUDGMENT. Was this a mere coincidence? Surely not! It was so designed by the ONE who inspired the Scriptures to show us the connection between the Law and Judgment. A broken LAW always brings JUDGMENT.

There were ELEVEN things that John saw in connection with the JUDGMENT at the Great White Throne. He saw (1) a great white throne; (2) Him that sat upon the throne; (3) the dead, small and great, stand before God; and (4) the books were opened, and (5) another book, which was the book of life; and (6) the dead were judged out of the things written in the books; and (7) the sea gave up its dead; and (8) death and hell delivered up their dead; and (9) these were judged, every man, according to their works; (10) he saw death and hell being cast into the lake of fire, and he saw (11) those cast into the lake of fire whose names were not found written in the book of life. (Revelation 20:11-15)

Let the reader read the passage referred to and see if these things are not listed correctly. Since John said, "They were JUDGED every man according to their works", in two places, the writer counted them and found ELEVEN things. (Read Revelation 20:12-13.) There was a reason for this repetition. It was perhaps to let us know that in this JUDGMENT the subjects will be those who are dead both spiritually and also physically, before being raised. This is the last JUDGMENT.

When the Bible talks about ELEVEN disciples, it means one has fallen away, JUDGMENT has fallen and one is under judgment.

It would be well for the reader to study the 24 times this number is used in the Bible.

TWELVE

GOVERNMENTAL PERFECTION

This number symbolizes God's perfect, divine accomplishment actively manifested. It shows a COMPLETENESS of a GROWTH or ADMINISTRATION. TWELVE marks GOVERNMENTAL PERFECTION and is used as the SIGNATURE of Israel. This number is used 187 times in the Bible. It is used 22 times in the Book of Revelation.

TWELVE is the number for GOVERNMENT BY DIVINE APPOINTMENT. Jesus said to His apostles,

> "Verily, I say unto you, that ye which have followed me, in the regeneration, when the Son of man shall sit in the throne of His glory, ye also shall sit upon TWELVE thrones, judging the TWELVE tribes of Israel." (Matthew 19:28)

The TWELVE Apostles shall sit upon their TWELVE thrones and judge and rule in connection with our Lord's rule upon His throne. They will occupy those TWELVE thrones by DIVINE APPOINTMENT.

Genesis 17:20 states that Ishmael begat TWELVE princes, and in Numbers 1:5-16 there are TWELVE princes named over the TWELVE tribes of Israel.

> "And, Solomon had TWELVE officers over all Israel, which provided victuals for the king and his household." (I Kings 4:7)

There are TWELVE months in the year. There are also TWELVE signs in the Zodiac. Dr. Seiss, in his "Gospel in the Stars", claims that the TWELVE signs of the Zodiac have been accepted by the astronomers throughout the centuries, and that none of them know where the mapping of the stars started. Some claim that it came from beyond the flood. In the opinion of Dr. Seiss, it was revealed by the Lord to such men as

Enoch and Noah. The TWELVE signs of the Zodiac were mentioned in the book of Job, the oldest book in the Bible. God said to Job,

> "Can you direct the signs of the Zodiac?" (Job 38:32) (In the King James Translation the word Mazzaroth means "The Twelve Signs".)

Even the testimony of the stars is numerically in harmony with the Bible. No wonder the Psalmist said,

> "The heavens declare the glory of God and the firmament showeth His handywork. Day unto day uttereth speech, and night unto night showeth knowledge." (Psalm 19:1-2)

There were TWELVE tribes of Israel to make up the nation of Israel. There were TWELVE stones in the High Priest's breastplate representing the nation of Israel. (Exodus 28:17-21) TWELVE cakes of shew-bread, one loaf for each of the TWELVE tribes were to be placed in the Holy Place. (Exodus 25:23-30 and Exodus 37:10-16; Leviticus 24:5-9)

TWELVE spies were sent out by Moses to spy out the land of Canaan. (Number 13:1-33)

When the children of Israel left the land of Egypt and started toward the land of promise, when they came to Elim, there they found TWELVE wells of water to quench their thirst.

> "And they came to Elim, where were TWELVE wells of water, and three score and ten palm trees: and they encamped there by the waters." (Exodus 15:27)

Elijah built an altar of TWELVE stones, and fire from heaven came down and consumed the altar offering. (I Kings 17:30-40)

Jesus chose TWELVE disciples to follow him. Jesus said, at His request, the Father would send TWELVE legions of angels. There was the woman who was diseased for TWELVE years. Jairus's daughter was TWELVE years of age when Jesus healed her. Jesus visited the Temple at TWELVE years of age.

In the Book of Revelation we read of the woman with a crown of TWELVE stars, and that the New Jerusalem has TWELVE gates, and at the gates TWELVE angels; and it has TWELVE foundations, and in them the names of the TWELVE Apostles of the Lamb; that its trees bear TWELVE manner of fruits; that it lieth four-square and measures TWELVE thousand furlongs on a side, and that the height of the wall is 144 cubits, or 12 x 12, which symbolizes this truth --, that there is nothing in God's city which is not perfected governed.

It is also significant that only TWELVE judges -- who judged Israel -- are recorded in the Book of Judges.

TWELVE is the number that comes next after ELEVEN. The reign of Christ and His Apostles and saints will follow the judgment of the false woman, (Revelation 17:1 to 19:2), and the JUDGMENT of the beast and his associates. (Revelation 19:11-21) The account of this reign is found in Revelation 20:1-6. This account follows immediately after the prophecy of the destruction of the beast, his armies, and the kings of the earth, as found in Revelation 19:11-21. TWELVE follows ELEVEN, and the reign of (Revelation 20:1-6 follows the JUDGMENT of Revelation 19: 11-21). How fitting and orderly it all is when one is in line with the truth!

Let the opponents of the Premillennial position; Anti-millennialists, Post-millennialists, and A-millennialists, take the numbers and prove their position if they can.

There are things more startling yet to come, before which the puny minds of men seem as nothing, and the greatness of our God eclipses all things else. In John 19:11 Jesus told Pilate that he could have no power, or authority against Him except it were given to him from above. This was expressed by the TWELVE Greek words: "Auk (1) eixes (2) exousian (3) audemian (4) kata (5) emou (6) ei (7) mn (8) an (9) soi (10) dedomenon (11) anothen (12)."

Here is the AUTHORITY that comes from above. If it comes from above, then it is AUTHORITY that comes from God. These TWELVE Greek words which Jesus used in telling Pilate that he could have no power, or authority against Him, unless it came from ABOVE, is exactly the number that has been found to stand for DIVINE AUTHORITY. Later on it shall be shown how this number fits in with the other numbers set forth.

THIRTEEN
DEPRAVITY AND REBELLION

The number THIRTEEN is associated with DEPRAVITY AND REBELLION. This number usually is in association with evil or badness. It is found some 15 times in the Bible. All the names of Satan are divisible by THIRTEEN. The name of reproach given to our Lord on the Cross "Jesus of Nazareth" is also divisible by THIRTEEN. THIRTEEN times the phrase, "Forever and ever" is used in the Book of Revelation.

Jesus mentions THIRTEEN things when He gave a picture of the REBELLIOUS and DEPRAVED heart of man, in Mark 7:21-22,

> "For from within, out of the heart of men, proceed (1) evil thoughts, (2) adulteries, (3) fornications, (4) murders, (5) thefts, (6) covetousness, (7) wickedness, (8) deceit, (9) lasciviousness, (10) an evil eye, (11) blasphemy, (12) pride, (13) foolishness."

If this is not a picture of a DEPRAVED heart, then where would one go to find it. Regardless of how much fallen man may resent this picture, it is the indictment, which the Son of God has spoken against him.

The word Dragon, which is a symbol of the Devil (Revelation 12:9), is found THIRTEEN times in the Book of Revelation. The Dragon, or Devil, is behind all REBELLION. (12:3; 12:4; 12:7 (Twice); 12:9; 12:13; 12:16; 12:17; 13:2; 13:4; 13:11; 16:13; 20:2)

Nimrod, in Genesis 10:6-8, was the THIRTEENTH in the line of Ham. The beginning of Nimrod's kingdom was Babel. (Genesis 10:9-10) It was at this place that men rebelled against the Lord's command to fill up the earth. (Genesis 11:1-9).

Genesis 14:4 says,

> "TWELVE years they served Chedorlaomer,

> and in the THIRTEENTH year they REBELLED."

Notice the number TWELVE associated with the reign or government of Chedorlaomer, and that the number THIRTEEN is associated with the REBELLION against that government. Here both numbers are used, TWELVE which represents the GOVERNMENT BY DIVINE AUTHORITY, and THIRTEEN representing REBELLION AGAINST GOD.

In Esther 3:8-13, Haman, the enemy of the Jews, had a decree signed on the THIRTEENTH day of the first month to have all the Jews put to death on the THIRTEENTH day of the TWELVETH month.

In Romans 1:29-31 there are TWENTY THREE things listed against sinful man. The THIRTEENTH is "Haters of God". How did it happen that the THIRTEENTH in the list was "Haters of God"? There can be but one answer: THIRTEEN is the number for the depraved, rebellious, sinful nature of man. Being depraved in his nature, he is a hater of God. "The carnal mind is enmity against God." (Romans 8:7). This is something for those to consider who deny the doctrine of DEPRAVITY.

Counting the Levitical tribe there were THIRTEEN divisions of that nation, the nation, Israel. Jacob had twelve sons. Their names were (1) Reuben, (2) Simeon, (3) Levi, (4) Judah, (5) Dan, (6) Naphtali, (7) Gad, (8) Asher, (9) Issachar, (10) Zebulun, (11) Joseph, and (12) Benjamin. (Genesis 29:32 to 30:24; 35:16-19)

Joseph had two sons, Manasseh and Ephraim. Jacob adopted them as his sons. (Genesis 48:5). This eliminated the name of Joseph from the list of the tribes, leaving ELEVEN. But it added the names of Ephraim and Manasseh, making THIRTEEN in all.

When the tribes were numbered in Number 1:1-46, the Levites were not numbered among them. "But the Levites after the tribe of their fathers were not numbered among them". That is why we so often read of the TWELVE tribes of Israel. But if the Levites are included in the count there are THIRTEEN tribes, or divisions.

There were TWELVE princes chosen for the TWELVE tribes that were numbered. The list of those tribes and their princes are found in Numbers 1:5-16. They are as follows:

	TRIBE	PRINCE
1.	Reuben	Elizur
2.	Simeon	Shelumiel
3.	Judah	Nahshon
4.	Issachar	Nathaneel
5.	Zebulun	Eliab
6.	Ephraim	Elishama
7.	Manasseh	Gamaliel
8.	Benjamin	Abidan
9.	Dan	Ahiezer
10.	Asher	Pagiel
11.	Gad	Eliasaph
12.	Naphtali	Ahira

If the reader will examine Numbers 1:5-16 he will find this list to be correct. If he will look closely he will find that the name of the tribe of Levi and the name of Aaron are not found in the list. When there was a REBELLION against the priesthood of Aaron (Numbers 16:1-47), the Lord had these TWELVE princes to bring their rods to Moses. Each man's name was to be put on his rod. Then Aaron brought his rod for the tribe of Levi, and it was put among the TWELVE rods. This made THIRTEEN rods in all, one for each of the twelve princes listed above, and one for Aaron. These rods were placed in the tabernacle over night. The next morning Aaron's rod was bearing (1) buds, (2) blossoms, and (3) almonds. They looked, and every man took his rod. Then God said to Moses,

> "Bring Aaron's rod again before the testimony, to be kept for a token against the REBELS." (Numbers 17:1-10)

Here the number THIRTEEN is connected with REBELLION.

In Aaron's rod there is also a picture of the resurrection of our Lord. When Aaron brought that rod to Moses it was a dead walking stick. The next morning it was a living stick with (1) buds, (2) blossoms, and (3) almonds. It was a witness against those who REBELLED and rejected the priesthood of Aaron. So also the resurrection of Christ was a witness against those who rejected Christ and had Him put to death. One can readily see that by reading such passages as Acts 2:32-37; 3:13-15; and 5:22-28.

There were THIRTEEN divisions of the promised land. While the Levites were not given any land there were two portions given to the tribe of Manasseh. Half of that tribe had a portion on the east side of Jordan. The other half of the tribe had its part on the west side of Jordan. The reader can verify this by referring to his Bible maps or by examining the books of Deuteronomy and Joshua.

In Deuteronomy 31:27 Moses said to Israel,

> "I know thy REBELLION, and thy stiff neck: behold, while I am yet alive with you this day, ye have been REBELLIOUS against the Lord: and how much more after my death?"

The THIRTEEN times their REBELLION in the wilderness is referred to symbolized their REBELLIOUS nature and conduct while Moses was yet alive. The THIRTEEN land divisions pictured their REBELLIOUS nature after the death of Moses. This is a picture of the REBELLIOUS nature of all mankind, both Jews and Gentiles, and of the Depraved nature of all mankind.

If the reader will examine Numbers 3:39-51 he will find that the Lord took the Levites in exchange for the firstborn. This exchange was made person for person until the number of the Levites was exhausted. Then there remained 273 of the firstborn who had to be redeemed with FIVE shekels per person. Divide 273 by 13, the number of DEPRAVITY, and the result is exactly TWENTY ONE, with nothing left over. The reader can make the division for himself. There are exactly TWENTY-ONE sins recorded against Israel from Egypt to Jordan. They are found in the following places:

1. Exodus 14:10-21	12. Numbers 12:1-15
2. Exodus 15:23-24	13. Numbers 14:1-11
3. Exodus 16:1-3	14. Numbers 14:40-45
4. Exodus 16:19-20	15. Numbers 15:32-36
5. Exodus 16:27-28	16. Numbers 16:1-35
6. Exodus 17:1-4	17. Numbers 16:41-50
7. Exodus 32:1-9	18. Numbers 20:1-6
8. Leviticus 10:1-2	19. Numbers 20:8-12
9. Leviticus 24:10-14	20. Numbers 21:4-9
10. Numbers 11:1-3	21. Numbers 25:1-9
11. Numbers 11:10-35	

The THIRTEENTH sin in the above list is where Israel REBELLED and REFUSED to go up and possess the land. (Numbers 14:6-9)

Israel's REBELLIOUS or DEPRAVED nature caused them to commit those TWENTY ONE sins. DEPRAVITY, which results in outward sins, makes necessary redemption for sin. The number for REBELLION, 13, multiplied by 21, the number of sins committed, equals 273, the number of firstborn that had to be redeemed after the exchange with the Levites. That redemption was done with FIVE shekels of silver per person. FIVE is the number for GRACE. (Ephesians 1:7) Silver is the symbol of Redemption.

> "In whom we have REDEMPTION through his blood, the forgiveness of sins, according to the riches of his GRACE."

Subtract 13, the number for DEPRAVITY or REBELLION, from 21, the number of sins they committed, and there are 8 left, the number for the NEW BIRTH. This is what Israel needed.

To make it still more marvelous, notice for a moment, the coverings of the Tabernacle. (Exodus 26:1-13) There are 10 linen curtains and 11 goat's hair curtains, making TWENTY ONE in all, the exact number of sins recorded against the children of Israel in the wilderness. Over these TWENTY ONE curtains was laid a covering of ram's skins dyed red, picturing death and blood. And beneath this covering were the TWENTY ONE curtains. Read Romans 4:7 and see the beauty and glory of this:

> "Blessed are they whose iniquities are forgiven, and whose sins are covered."

God was saying to Israel, and to us, that the shed blood of Christ is a covering for all sins.

> "O the depth of the riches both of the wisdom and knowledge of God! How unsearchable are His judgments, and His ways past finding out." (Romans 11:33)

The shed blood shelters not only from sins already committed, but from all future sins also, and provides redemption for all future time.

Can another such book be found as the Word of God? Can such as this be found in the book of Mormon, the Koran, the writings of philosophers, or other religious leaders? Can they show such foreknowledge? Can they look into the future and make it out with figures as has the Word of God? Let the doubter answer if he can.

FOURTEEN
DELIVERANCE OR SALVATION

FOURTEEN is the number that represents DELIVERANCE or SALVATION. It is used some 26 times in the Bible. It was the FOURTEENTH day of the first month of the year when the children of Israel were DELIVERED from Egyptian bondage, and from the stroke of judgment which fell upon the firstborn of the Egyptians. (Exodus 12:6-7; Exodus 12:12-13; Leviticus 23:4-5)

The number FOURTEEN is found three times over connected with Christ's coming into the world, and He came to SAVE, or DELIVER His people from their sins.

> "So all the generations from Abraham to David are FOURTEEN generations; and from David until the carrying away into Babylon are FOURTEEN generations, and from the carrying away into Babylon unto Christ are FOURTEEN generations. (Matthew 1:17)

The Salvation of the soul takes place when one believes, at which time, he comes under the blood of Christ, "Our Passover". But the body will not be delivered from the bondage of corruption until the resurrection of the believer or the rapture of the saints, then the redemption of our body takes place. (Romans 8:20-23)

This explains why the numbers FOURTEEN and THREE are found together. FOURTEEN is for DELIVERANCE and THREE is for the RESURRECTION. In I Chronicles 25:4-6 there are mentioned FOURTEEN sons of Heman, and THREE daughters who were for singers in the house of the Lord. Israel was DELIVERED from the plague in Egypt on the FOURTEENTH day. THREE days later they passed through the Red Sea, where is a figure of the RESUR-

RECTION. Then they sang a song unto the Lord. In that song, they said, "The Lord is my strength and song, and He is become my SALVATION." (Exodus 15:1-2) This is in line with the FOURTEEN sons and THREE daughters of Heman who were singers in the house of the Lord. The children of God rejoice in the SALVATION of their souls, and sing for joy. And they rejoice in the hope of their RESURRECTION.

On his voyage to Rome, Paul and his company were caught in a violent storm. (Acts 27:14-44) When the men had despaired of any being saved God sent His angel and told Paul that they would all escape alive. Upon the authority of God's Word Paul told them none would die. (vs 22-25) There were 276 on board the ship. (vs 37) In Romans 1:29-32 we learn that TWENTY THREE is the number of Death. TWELVE, the number for Divine Authority, goes into 276 exactly TWENTY THREE times. 276 divided by 12 equals 23. They were saved from death in the storm on the FOURTEENTH day. (vs 33-44) This is in keeping with the number FOURTEEN being the number of Salvation.

In Galatians 1:15-16, Paul tells about God revealing His Son in him that he might preach Him among the Gentiles. A few verses below he tells that FOURTEEN years after this he went up to Jerusalem with Barnabas and Titus. (Galatians 2:1) The record shows that Paul's mission on this occasion was to confer with the apostles and elders as to the question of circumcision being necessary to SALVATION. (Acts 15:1-12) Paul's name occurs the FOURTEENTH time in this connection. If one begins to count where it said,

> "Then Saul, (who is also called Paul) etc." (Acts 13:9)

then the FOURTEENTH time Paul's name is found is in Acts 15:2 where it is said,

> "They determined that Paul and Barnabas, and certain other of them, should go up to Jerusalem unto the apostles about this question."

The question was about circumcision being essential to Salvation. But if the count commences in Acts 13:13 when Paul is no longer called Saul, but Paul only, then the FOURTEENTH time Paul's name is found is in Acts 15:12, just after Peter had said,

> "We believe that through the grace of our Lord Jesus Christ we shall be SAVED, even as they," (vs 11)

In either case Paul's name is found the FOURTEETH time in connection with the discussion about SALVATION. And Galatians 2:1-5 shows that this was FOURTEEN years after Paul himself was SAVED.

FIFTEEN
REST

FIFTEEN is the number for REST. REST is the result of DELIVERANCE, or SALVATION. Three times in the year Israel RESTED on the FIFTEENTH day of the month.

> "On the FIFTEENTH day of the same month (the first month) is the feast of the unleavened bread unto the Lord: seven days ye must eat unleavened bread. In the first day (15th) ye shall have an holy convocation: ye shall do no servile work therein." (Leviticus 23:6-7)

The second time the people rested was the FIFTEENTH day of the Seven Month.

> "The FIFTEENTH day of this SEVENTH month shall be the feast of the tabernacles for seven days unto the Lord. On the first day shall be an holy convocation: ye shall do no servile work therein." (Leviticus 23:34-35)

On these two occasions they RESTED from their work on the FIFTEENTH day.

The third time is found in the book of Esther.

> "And Mordecai wrote these things, and sent letters unto all the Jews that were in the provinces of King Ahasuerus, both nigh and far, to establish this among them that they should keep the fourteenth day of the month Adar, and the FIFTEENTH day of the same, yearly, as the days wherein the Jews RESTED from their enemies, etc." (Esther 9:20-22)

The FIFTEENTH time the name of Naomi is found is where she said to Ruth,

> "My daughter shall I not seek REST for thee?" (Ruth 3:1)

That rest came through Boaz their kinsman redeemer. (Chapter 4) The word KINSMAN is found FIFTEEN times in the book of Ruth.

After the first creation God RESTED on the SEVENTH day. Add EIGHT, the number for the NEW BIRTH, to SEVEN, the day God rested from the first creation, and you get FIFTEEN, the number for spiritual REST. SEVEN plus EIGHT equals FIFTEEN.

Perhaps the reader has observed that the study of Bible numbers, and their application to Bible truths, is like learning mathematics in the life of a child. He must learn the smaller numbers before he is ready to advance in his study of arithmetic. So our knowledge of the smaller Bible numbers helps us in our understanding of the larger ones.

This number, FIFTEEN, is found TWENTY FOUR times in the Bible.

SIXTEEN

LOVE

SIXTEEN is the number that represents LOVE. This number is found 23 times in the Bible. There are SIXTEEN times in the Bible we find JEHOVAH TITLES. The word JEHOVAH means the Self-existent or Eternal Creator; the Immutable One, He who WAS, and IS, and IS TO COME. The name JEHOVAH is combined with other words which form what we know as the JEHOVAH TITLES.

1. Jehovah - Elohim — the Eternal One or Creator. (Genesis 2:4-25)
2. Adonai-Jehovah - the Lord our Sovereign; Master Jehovah. (Genesis 15:2,8)
3. Jehovah - Jireh — the Lord will see or provide. (Genesis 22:8-14)
4. Jehovah - Nissi — the Lord our banner. (Exodus 17:15)
5. Jehovah - Ropheka — the Lord our healer. (Exodus 15:26)
6. Jehovah - Shalom — the Lord our peace. (Judges 6:24)
7. Jehovah - Tsidkeenu — the Lord our righteousness. (Jeremiah 23:6)
8. Jehovah - Mekaddishkem — the Lord our sanctifier. (Exodus 31:13; Leviticus 20:8; 21:8)
9. Jehovah - Saboath — the Lord of hosts. (I Samuel 1:3. Used 281 times)
10. Jehovah - Shammah — the Lord is present. (Ezekiel 48:35)
11. Jehovah - Elyon — the Lord most high. (Psalm 7:17; 47:2; 97:9)
12. Jehovah - Rohi — the Lord my shepherd. (Psalm 23:1)
13. Jehovah - Hoseenu — the Lord our Maker. (Psalm 95:6)
14. Jehovah - Eloheenu — the Lord our God. (Psalm 99:5; 8, 9)

15. Jehovah - Eloheka — the Lord thy God. (Exodus 20:2; 5, 7)
16. Jehovah - Elohay — the Lord my God. (Zechariah 14:5)

In I Corinthians 13:4-8 there are SIXTEEN things said about LOVE. The SIXTEENTH time Paul's name occurs is where he is called "BELOVED". (Acts 15:25)

There were EIGHT boards and SIXTEEN sockets in the west side of the TABERNACLE. (Exodus 26:25) In the EIGHT boards is given the number for the NEW BIRTH. In the SIXTEEN sockets under those EIGHT boards is the number for LOVE.

> "Everyone that LOVETH, is born of God, and knoweth God." (I John 4:7)

SIXTEEN is two times EIGHT. The one who is BORN AGAIN (EIGHT) loves all other persons who are BORN of God (EIGHT).

> "Everyone that LOVETH Him that begat LOVETH Him also that is begotten of Him." (I John 5:1)

How beautifully and perfectly these numbers fit into the whole pattern of the WORD of GOD.

SEVENTEEN

VICTORY

In this number we get an example of how the Holy Spirit uses more advanced arithmetic to bring out spiritual truths. SEVENTEEN is a very well-marked number in Scripture. It is the SEVENTH prime or indivisible number, and therefore we might expect that it would bear a special significance in God's Word. THIRTEEN is the SIXTH prime or indivisible number, and just as SIX is the number of man, and especially of man without God, glorying in his own strength, so the SIXTH prime number, THIRTEEN, shows the result of man following the imaginations of his own heart, that is, sin, rebellion against God's perfect, holy Will. So, SEVENTEEN, being the SEVENTH prime number, and SEVEN is the number of completeness and spiritual perfection, we should expect to find SEVENTEEN bearing some relation to that meaning, and that is exactly what we do find, for all through God's Word, both the Old and New Testaments, SEVENTEEN is revealed as bearing the significance of VICTORY, complete VICTORY in Christ Jesus.

It is a number usually compounded in Scripture of the numbers SEVEN and TEN. Now SEVEN signifies COMPLETENESS and TEN signifies TESTIMONY, so the two combined have the deeper significance of the TESTIMONY OF COMPLETENESS or VICTORY that is found in the Lord.

We find the use of SEVENTEEN in the seemingly dry details of the patriarchs' ages in Genesis the 5th chapter. Here we have recorded the oldest man who ever lived, the son of a man who "walked with God", and therefore presumably the godly son of a saintly father, we can divide his age by SEVENTEEN, and it will leave no remainder. 969 is a multiple of SEVENTEEN the complete age of Methuselah. When he was 187 years of age he begat Lamech which also divides by SEVENTEEN. Of none of the other patriarchs is this true, the only other age in the chapter that will so divide is Lamech's age after the birth of his son, Noah, another antediluvian who "walked with God", and Lamech's age was 182 and he lived 595 years more, and this is a multiple of SEVENTEEN.

Now Methuselah's name means, "When he is dead it shall be sent," and a little calculation will show that he died just before the Flood, so God in the ages of these two men, one, Methuselah, to be the one whose death should be the signal of the dread judgment to come, and the other, Lamech, of whom was born, Noah, the godly man, to be preserved through the Flood, marked with this significant number SEVENTEEN -- VICTORY, the fact that in spite of all the accumulating wickedness of man, He purpose to provide VICTORY in the lives of the believers. Here is beautifully portrayed, in the midst of the stormy waters of judgment, God's Ark of safety, a means of salvation to Noah and his family. Truly this is VICTORY for mankind.

Then among the antediluvian patriarchs we only read of two, who are said to have "walked with God," one, Enoch, was raptured to heaven, the other preserved in the Ark through the waters of judgment. Now Enoch was "the SEVENTH from Adam", and Noah was the TENTH, so here again we get the VICTORY brought out by these two numbers added together making SEVENTEEN.

Further in the actual story of the Flood, not only ages of men and order of descent, but calendar dates are made use of to bring out this truth. The day the Ark floated safely on the rising waters, when the fountains of the great deep were broken up, was the "SEVENTEENTH day" of the second month, and the day it came safely to rest on "the mountains of Ararat" was the same "SEVENTEENTH day" but of the SEVENTH month! (Read Genesis 7:11 and 8:4)

> "In the six hundredth year of Noah's life, in the second month, the seventeenth day of the month, the same day were all the fountains of the great deep broken up, and

the windows of heaven were opened. (Genesis 7:11)

"And the ark rested in the seventh month, on the seventeenth day of the month, upon the mountains of Ararat." (Genesis 8:4)

It was on this very same day, the SEVENTEENTH day of the SEVENTH month, that, years later, God again showed His mighty power. After Israel went out of Egypt on the FOURTEENTH day of the month they made THREE days journey and crossed the Red Sea. (See Exodus 8:27; 12:-1-13; 12:37; 13;20; 14: 1-2) Thus in the night of the SEVENTEENTH day of the month they made the passage of the Red Sea making a picture of the Resurrection. When they came out of that sea at the break of day they sang the song of VICTORY. (Exodus 15:1-21) In the opening of that song they said,

"I will sing unto the Lord, for He hath TRIUMPHED gloriously." (Verse 1)

In that day when the saints of God shall have been made VICTORIOUS over the grave they shall sing God's praise and shout,

"O death where is thy sting? O grave where is thy VICTORY? Thanks be unto God which giveth us the VICTORY through our Lord Jesus Christ." (I Corinthians 15: 55-57)

To mark the spiritual truth that lay behind this wonderful salvation, redemption by blood and power, God bade them alter the month from SEVENTH to FIRST.

"This month shall be unto you the beginning of months: it shall be the first month of the year to you." (Exodus 12:2)

Thus did God early in His Word emphasize the same truth that the Lord Jesus brought home to Nicodemus' heart, "Ye must be born again." A new birth, a new beginning, is essential to all who would escape from the judgment to come.

Yet a third time was this significant day chosen, to fulfil the pictures drawn years before by the hand of the Holy Spirit. And on the SEVENTEENTH day of the month, now the first month of the Jewish sacred year, did God the Father break the bars of the tomb, and "for our justification" raise His only Begotten Son from the dead, "now no more to return to corruption."

Our Lord was crucified on the occasion of the celebration of the Passover, which came on the FOURTEENTH day of the month. (Leviticus 23:5) Luke tells about our Lord eating the Passover with His disciples the same night that He instituted His memorial supper. (Luke 22:13-20) He was taken that night after Judas betrayed Him before the High Priest. The next morning He was taken before Pilate and was delivered up to be crucified. (Luke 22:47 to 23:33) Remembering that God reckoned time from one evening to the next evening (Genesis 1:5), then it was still the FOURTEENTH day of the month when Christ was the Lamb slain on the altar. While the Jews were observing the Passover, Christ the Passover Lamb was already slain and in the tomb. He was dead THREE days and nights. (Matthew 12:39-40) These THREE days added to the FOURTEENTH day would bring it to the SEVENTEENTH day of the month, which was the day of the FIRST FRUIT Offering which took place on the FIRST day of the week, the day of the Resurrection, in which our Lord became VICTORIOUS over death and the grave.

Well may we rejoice then at the glorious truth taught by this number, SEVENTEEN. Our very hope of a joyful resurrection shines through it, for "if Christ be not raised, your faith is vain; ye are yet in your sins." (I Corinthians 15:17) It is in virtue of our belief in what happened on that glad SEVENTEENTH day, that we "rejoice in hope of the glory of God."

Among the Old Testament types of the Lord Jesus, by far the fullest and most outstanding is Joseph, and very strikingly does this number lend its aid to bring out this fact. We read in Genesis 37:2 that Joseph was "SEVENTEEN years old" when his father, Jacob, sent him out of the vale of Hebron to Shechem to "see the peace of thy brethren" (verse 14).

> "These are the generations of Jacob. Joseph, being seventeen years old, was feeding the flock with his brethren; and the lad was with the sons of Bilhah, and with the sons of Zilpah, his father's wives: and Joseph brought unto his father their evil report." (Genesis 37:2)

> "And he said to him, Go, I pray thee, see whether it be well with thy brethren, and well with the flocks; and bring me word again. So he sent him out of the vale of Hebron, and he came to Shechem." (Genesis 37:14)

Here we have a type picture of the Lord Jesus leaving heaven to do the Father's Will, and coming down to earth to His brethren according to the flesh, the Jewish nation. He was rejected, as Joseph was, at this His First Coming. But presently He is coming again, and this time it will be to rule the whole world, and His brethren, the Jews, will be reconciled to Him. So when next Jacob sees his son Joseph again, it is to find him "lord of all Egypt," and when he goes down to see him, we read,

> "and Jacob lived in the land of Egypt SEVENTEEN years." (Genesis 47:28)

Thus, are both the type-pictures, in Joseph's life, of Christ's First Coming, and His Second Coming, marked by the figure-SEVENTEEN, to bring out the truth that God is over-ruling all things to bring about VICTORY in the earth! Jacob sees Joseph for SEVENTEEN years at the beginning of Joseph's life, and then again he sees him for another SEVENTEEN years before he dies. Each period pointing, one to the FIRST, and the other to the SECOND COMING of the Lord Jesus Christ, and stamping both Comings as revealing the VICTORY that comes in order, first, "the Sufferings," and then "the Glory that should follow." Both needful to free this poor earth from the hateful effects of sin.

In the last chapter of Joshua there is a remarkable address made to the tribes of Israel at Shechem, who are at this time settled safely in the Promised Land. In this address which Joshua repeats from the mouth of "the Lord God of Israel", God recounts all that His mighty power has accomplished for the children of Israel from the days of Abraham to the present time, and in it there is a repetition of exactly SEVENTEEN "I's" beginning with Verse 3,

> "And I took your father Abraham from the other side of the flood,"

and closing with, verse 13,

> "and I have given you a Land for which ye did not labour,"

thus bringing out that the VICTORY that had been gained was in all God's wondrous dealings with His chosen people. It was all of God and none of man's labour.

You will find the same number of exactly SEVENTEEN "I's" in Ezekiel 16:6-14 describing God's blessing on Jerusalem. Their blessings and VICTORY is from God.

When we turn onward to the book of Psalms, we find that the name of the Holy City, Jerusalem, occurs just SEVENTEEN times in the Psalms! While the third or Leviticus book of Psalms, that is specially the book of Worship, has exactly SEVENTEEN Psalms (73-89), and the fourth or Numbers book, the Wilderness book, has also SEVENTEEN Psalms only (90-106). Thus is this great truth again brought out, that the VICTORY is WORSHIP and WALK is learn through the study of God's Word.

When we move forward to the Prophets, we still

find the same significance attaching to this number. Doubtless many have noticed when reading Isaiah and Jeremiah, that while the former is never recorded as praying, Jeremiah's first chapter records a prayer, and from thence Jeremiah constantly records his prayers to God, until the 32nd chapter is reached, and then it will have been noticed that his prayers suddenly cease, and though his book continues through another TWENTY chapters, yet never a prayer is found therein. Why this sudden silence? Perhaps the "higher" critics, who have already in the foolish imagination of their hearts, supposed a second or deutero-Isaiah to have written the evangelistic portion of that book, would find evidence in this cessation of prayers, of a second Jeremiah, who composed the last half of that book!

The reason of this absence of prayers in the latter half of Jeremiah is a part of the evidence of the Holy Spirit's Authorship of the whole Word of God.

The prayers of Jeremiah will be found to number exactly SEVENTEEN! And when we come to the 32nd chapter, which contains the SEVENTEENTH prayer, we find a remarkable scene. The city of Jerusalem is closely invested by the army of Nebuchadnezzar, and God has already said, through His prophet Jeremiah, that the city is to fall into the hands of the Chaldeans. Jeremiah is a close prisoner "in the court of the prison which was in the king of Judah's house." (Jeremiah 32:2) But God's Word comes to him that his cousin, Hanameel, is going to come to him, to ask him to buy his field in their native village of Anathoth (which means "Answered Prayers"), a few miles north of Jerusalem.

Now seeing that the whole country was in the hands of the Chaldeans, and there was apparently no chance of taking possession of the property, it would seem a very foolish way of spending money. Evidently Jeremiah thought so too, and perhaps wondered whether it could really be of God's Word, but Hanameel duly comes to him and puts this curious request, and Jeremiah says,

> "Then I knew that this was the word of the Lord." (Jeremiah 32:8)

So, prisoner though he was, he bought the field,"

> "and weighed him the money, even SEVENTEEN shekels of silver," (Jeremiah 32:9)

Jeremiah has read his Bible to some purpose, and has learned his lesson, and having paid the significant purchase price for a piece of land that presumably he would never possess, he drops upon his knees and prays his SEVENTEENTH prayer!

It begins in the SEVENTEENTH verse, and in it he tells God all his perplexity and wonder at being told to do such a seemingly foolish thing as to pay SEVENTEEN shekels for a field even then in the possession of an invading enemy. Then God answers him, and shows him the vision of the present time in which we live, the time when God would gather His earthly people out of all the countries where He had driven them, and would bring them again to the Holy Land and cause them to dwell safely,

> "and they shall be My people, and I will be their God,"

and fields again should be bought and sold for money, and they should again be planted in the land which God had given them. No further prayer is recorded in the book, evidently to show us that our prayers will end with the Millennium, when faith will be lost in sight, and Christ will be visibly present in the midst of His saints reigning over the earth. (Jeremiah 32:37-44)

When we turn to the New Testament we find still the same significant use of this number SEVENTEEN. We have already mentioned that our Lord's Resurrection took place on the SEVENTEENTH day of the FIRST sacred month. And when we come to the Day of Pentecost we find the number again shining out only in a different way. For Acts 2:9, 10, 11 tells us that SEVENTEEN tongues were spoken on that great day that marked the Coming of the Holy Spirit from heaven!

In Romans 8:35-39 there are SEVENTEEN things listed which are unable to "separate us from the love of Christ." These are divided again into TWO lists of SEVEN and TEN to bring out the full meaning of the number.

> "Who shall separate us from the love of Christ? shall (1) tribulation, (2) or distress, (3) or persecution, (4) or famine, (5) or nakedness, (6) or peril, (7) or sword? etc. . . . . (38 vs.) For I am persuaded, that neither (8) death, (9) nor life, (10) nor angels, (11) nor principalities, (12) nor powers, (13) nor things present, (14) nor things to come, (15) nor height, (16) nor depth, (17) nor any other creature, shall be able to separate us from the love of God, which is in Christ Jesus our Lord." (Romans 8:35-39)

In I Peter will be found a list of things "of God". These two words are mentioned exactly SEVENTEEN times. Take your Bible and look up "of God". It is amazing!

Words only used SEVENTEEN times by the Holy Spirit are generally of great significance, for instance: "CHARISMA," which means "GIFT" is always spoken of God's gifts, its gematria too is 952, SEVENTEEN times SEVEN times EIGHT.

"KETHAB", meaning "SCRIPTURE" occurs SEVENTEEN times in the Old Testament.

"SEMEION", meaning "SIGN," often translated miracle, occurs SEVENTEEN times in John's Gospel. All the EIGHT miracles in that Gospel are signs, that is, they have a hidden typical meaning behind them.

"AGAPE", meaning "LOVE", occurs SEVENTEEN times in the first epistle of John.

"APHESIS", meaning "REMISSION", occurs only SEVENTEEN times in the New Testament. A study of the passages is replete with interest, and shows us that remission or putting away of sins is impossible without "shedding of blood", so that there is no hope for those who refuse to shelter under God's Red-Cross flag.

The SEVENTEENTH time the word "World" occurs in I John is where it is said,

> "Whatsoever is born of God OVERCOMETH the world." (I John 5:4)

This writer has found SEVENTEEN places where Christ is said to be at the right hand of God, of the Father, or at the right hand of power.

In Revelation 3:21, Jesus said,

> "To him that overcometh will I grant to sit with me on my throne, even as I also OVERCAME, and am set down with my Father in His throne."

We might, perhaps, however, conclude this chapter by showing how the constituent numbers of SEVENTEEN are sometimes used in such a manner, that, while separated, they yet suggest the fact that the passages in which they occur are intended to be read together, to be as it were added to one another, so that the significance of both the numbers SEVEN and TEN may bring out the full meaning of SEVENTEEN, such as we have already noticed in the case of Enoch and Noah.

There is for instance, a phrase only used twice by the Holy Spirit, and in one case the number SEVEN is inserted, in the other the number TEN. In Ruth 4:15 occurs the phrase,

> "better to thee than SEVEN sons,"

and in I Samuel 1:8 it occurs for the second time, only here it reads,

> "better to thee than TEN sons."

Now these passages occur in separate books, and yet in

consecutive chapters, one is in the last chapter of Ruth, the other is in the first chapter of 1st Samuel, which follows Ruth! The stories are entirely different, and yet both are typical of the Time of the End, when the Lord Jesus Christ, the heavenly Lord of the harvest, will take unto Himself His Bride.

Thus does SEVENTEEN again point forward to that glad time when VICTORY will be established on earth!

This number SEVENTEEN is found only TEN times in the Bible.

EIGHTEEN
BONDAGE

EIGHTEEN is the number that stands for BONDAGE. In Luke 13:16 Jesus said,

> "Ought not this woman, being a daughter of Abraham, whom Satan hath BOUND, lo, these EIGHTEEN years, be loosed from this BOND on the Sabbath day?"

Jesus said, "Whosoever committeth sin is the servant of sin." (John 8:34). There were EIGHTEEN sinners, or people, who were in BONDAGE to sin in Luke 13:4-5.

> "Those EIGHTEEN upon whom the tower in Siloam fell, and slew them, think ye that they were sinners above all men that dwell in Jerusalem? I tell ye, Nay: but, except ye repent, ye shall all likewise perish."

On two different occasions in the Book of Judges, the children of Israel are found in BONDAGE to their enemies EIGHTEEN years.

> "So the children of Israel served Eglon the king of Moab EIGHTEEN years." (Judges 3:14)

> "And the anger of the Lord was hot against Israel, and He sold them into the hands of the Philistines, and into the hands of the children of Ammon. And that year they vexed and oppressed the children of Israel: EIGHTEEN years, all the children of Israel that were on the other side of Jordan, in the land of the Amorites, which is in Gilead." (Judges 10:7-8)

Stephen said,

> "And God spake on this wise, That His seed

> (Abraham's) should sojourn in a strange land; and that they should bring them into BONDAGE, and entreat them evil four hundred years." (Acts 7:6)

Stephen is quoting from Genesis 15:13.

In the Old Testament there are EIGHTEEN places where this BONDAGE is spoken about. They are found in the following:

1. Genesis 15:13-14	10. Exodus 20:2
2. Exodus 1:14	11. Deut. 5:6
3. Exodus 2:23	12. Deut. 6:12
4. Exodus 2:23	13. Deut. 8:14
5. Exodus 6:5	14. Deut. 13:5
6. Exodus 6:6	15. Deut. 13:10
7. Exodus 6:9	16. Deut. 26:6
8. Exodus 13:3	17. Joshua 24:17
9. Exodus 13:14	18. Judges 6:8

By checking the above list it is found that the FIFTH time the word BONDAGE occurs in Exodus 6:5. The covenant referred to here was the covenant made with Abraham, Isaac and Jacob. (vs 3) According to Genesis 15:17-18, Abraham offered FIVE sacrifices the day God made the covenant with him. Romans 4:13-16 states that the promise to Abraham and his seed was through the righteousness of faith, and by faith, that it might be by GRACE. So the FIFTH time the word BONDAGE occurs is where God says He remembered the covenant made with Abraham, and He announced His purpose of redeeming Israel from BONDAGE.

By checking the above list the reader will also see that the TEN place, BONDAGE is mentioned in Exodus 20:2. Read that verse and the next one.

> "I am the Lord thy God, which have brought thee out of the land of Egypt, out of the house of BONDAGE. Thou shalt have no other gods before me."

Notice that the first word of the Ten Commandments

is the word after the word BONDAGE. It will also be noticed that it is the TENTH time Israel's Egyptian BONDAGE is mentioned. Paul calls the law the yoke of BONDAGE.

> "Stand fast therefore in the liberty wherewith Christ hath made us free, and be not entangled in the yoke of BONDAGE." (Galatians 5:1)

Only God could have arranged it so the TENTH time Israel's BONDAGE was mentioned would be followed by the first word in the TEN Commandments (or law), which is called the yoke of BONDAGE.

The number THIRTEEN stands for a rebellious, sinful, depraved heart and nature. In Mark 7:21-22, Jesus mentioned THIRTEEN evil things that came out of the heart of man. The THIRTEENTH time the word BONDAGE is found in the above list is in Deuteronomy 8:14. In that verse there is a warning to the Israelites that their heart be not lifted up.

> "Then thine HEART BE LIFTED UP, and thou forget the Lord thy God, which brought thee forth out of the land of Egypt, from the house of BONDAGE."

So the THIRTEENTH time the word BONDAGE occurs is in a place where Israel is warned against a REBELLIOUS heart.

Let the reader consider the SEVENTEENTH time the word BONDAGE occurs in the above list, remembering that SEVENTEEN is the number that stands for VICTORY. This is found in Joshua 24:17 in which connection Joshua mentions Israel's VICTORY over all her enemies.

> "For the Lord our God, He it is that brought us up and our fathers out of the land of Egypt, from the house of BONDAGE, and which did those great signs in our sight, and preserved us in all the way we went,

> and among all the people through whom we passed: and the Lord drave out from before us all the people." (Joshua 24:17-18)

This describes VICTORY for Israel over all her enemies, and occurs in connection with the SEVENTEENTH time Israel's BONDAGE is mentioned.

Nothing short of Divine Wisdom could arrange things like this. This number is used TWENTY-TWO times in the Bible.

The EIGHTEENTH time the word BONDAGE is found in the above list is in Judges 6:8.

> "The Lord sent a prophet unto the children of Israel, which said unto them, Thus saith the Lord God of Israel, I brought you up from Egypt, and brought you forth out of the house of BONDAGE . . . And I said unto you, I am the Lord your God; fear not the gods of the Amorites, in whose land ye dwell: but ye have not obeyed my voice." (Verse 10)

At the time the prophet spoke these words the Israelites were in BONDAGE to the Midianites. (Judges 6:1-11) This is evidence that the number EIGHTEEN stands for BONDAGE. The EIGHTEENTH and last time Israel's Egyptian BONDAGE is referred to was at the time they were in BONDAGE to the Midianites.

The EIGHTEENTH time Israel's bondage is referred to, which was in this place, completes the full number of times their Egyptian BONDAGE is mentioned. It came at a time when the Midianites had them in BONDAGE. Read now Judges 6:1,

> "The children of Israel did evil in the sight of the Lord: and the Lord delivered them into the hand of Midian SEVEN years."

Since SEVEN stands for COMPLETENESS, then the time Israel was in BONDAGE to Midian SEVEN years completed the EIGHTEEN times their Egyptian

BONDAGE is mentioned. This last reference COMPLETED the number (EIGHTEEN) that stands for BONDAGE.

In addition to the word BONDAGE being used EIGHTEEN times concerning Israel's experience in Egypt, the word BONDMEN is used FOUR TIMES. FOUR is the number for the UNSAVED man. This shows that man in the flesh is in BONDAGE to sin. (Galatians 4:3 and John 8:34-36)

NINETEEN
FAITH

NINETEEN is the number of FAITH. There are NINETEEN different persons referred to in Hebrew 11:1-32. While the name of Joshua is not mentioned in verse 30, yet Joshua was the leader of Israel at the overthrow of Jericho. The list is as follows: Through FAITH

1. We - Hebrew 11:3
2. Abel - Verse 4
3. Enoch - Verse 5
4. Noah - Verse 7
5. Abraham - Verse 8-10 and 17-19
6. Sarah - Verse 11
7. Isaac - Verse 20
8. Jacob - Verse 21
9. Joseph - Verse 22
10. Moses - Verse 23, 29
11. Joshua - Verse 30
12. Rahab - Verse 31
13. Gideon - Verse 32
14. Barak - Verse 32
15. Samson - Verse 32
16. Jephthae - Verse 32
17. David - Verse 32
18. Samuel - Verse 32
19. The Prophets - Vs. 32

Now, look at this and at the number under which each occurs. Moses is the TENTH. TEN is the number that represents the LAW, and "The LAW was given by Moses." (John 1:17). Abel is found in number TWO, which is the number for DIVISION. It is said that he offered a more excellent sacrifice than Cain. This led to division between him and Cain. Enoch is found in number THREE, the number for RESURRECTION, the Divine Number. Enoch's translation foreshadows the translation of the living saints at the rapture. Noah is the FOURTH mentioned and number FOUR is the number for the WORLD, which was destroyed in Noah's time. Abraham is listed under number FIVE, which represents GRACE. He and his seed are to be heirs by grace.

Jacob who is number EIGHT, which is the number for the NEW BIRTH, was the second born of twins. This represents the SECOND BIRTH, or the NEW

BIRTH. The name of Joseph is found in number NINE, representing the FRUIT OF THE SPIRIT. (Galatians 5:22-23) Jacob prophesied of Joseph and said,

> "Joseph is a fruitful bough, even a fruitful bough by a well, whose branches run over the wall." (Genesis 49:22)

Jesus taught the woman at Jacob's well about the well of water that would spring up into everlasting life. (John 4:5-14) Verse 5 says that Jacob gave this parcel of ground to Joseph. Joseph's name follows Jacob's in Hebrew 11:21-22. NINE follows EIGHT. The FRUIT OF THE SPIRIT follows the NEW BIRTH. Joseph's name follows Jacob's in John 4:5 and the well is mentioned next in verse 6. It was at this well Jesus instructed the Samaritan woman about the WELL of water that brings everlasting life. Just before this, Jesus had talked to Nicodemus about the NEW BIRTH, the birth of water and the Spirit. Now read all this in connection with the statement that "Joseph is a FRUITFUL bough by a well," and see how it ties together.

David is the SEVENTEENTH listed in the heroes of faith. This is the number for VICTORY. David was victorious over all his enemies. (II Samuel 7:1)

In Hebrew 11: verse 3, it is said,

> "Through faith we UNDERSTAND that the worlds were framed by the word of God."

So it is through faith that we UNDERSTAND. In Romans 1:29-31 Paul charges sinful men with TWENTY THREE things. The NINETEENTH charge he brings against them is being "without UNDERSTANDING." Being without FAITH, which is represented by the number NINETEEN, they are without UNDERSTANDING, for it is by FAITH that men understand.

In Paul's discussion of justification by FAITH in Romans 3:21 to 5:2 he used the word "FAITH" NINETEEN times. The reader may check on this for himself. The word FAITH is not found again in the

Book of Romans, until Romans 9:30 when Paul is contrasting Israel's blindness with the FAITH of the Gentiles.

The curtains of the court of the Tabernacle were 100 cubits long on both the north and south sides, and FIVE cubits high. (Exodus 27:9-18) FIVE represents GRACE. There were 20 pillars in each of these long sides. (vs 10) This gives NINETEEN spaces between the twenty pillars, and connects the number NINETEEN with the number FIVE, and Ephesians 2:8 says, "By grace (5) are ye saved (14) through faith (19): and that not of yourselves: it is the gift of God."

This number is found only three times in the Bible.

TWENTY
REDEMPTION

The number TWENTY is found 288 times in the Bible and it stands for REDEMPTION. The males of the children of Israel had to offer a ransom for their souls at the age of TWENTY. In connection with this ransom TWENTY gerahs are mentioned. (Read Exodus 30:12-14). The money that was given for their RANSOM was SILVER money. SILVER is a symbol of REDEMPTION.

SILVER money was used to make the 100 silver sockets in the Tabernacle, and the SILVER fillets and hooks for the pillars of the court. The reader will find that in Exodus 38:25-28. This connects the number Twenty with REDEMPTION and the SILVER with REDEMPTION. There were TWENTY boards on each side of the TABERNACLE north and south. (Exodus 26:18-19) The number TWENTY is used in describing the boards for each side, both north and south. The SILVER sockets were also mentioned in connection with each side. This shows forth a two-fold REDEMPTION, a REDEMPTION for the body, and a REDEMPTION for the soul.

The same thing is pictured in the TWENTY pillars with their SILVER fillets and hooks on the south and north sides of the court of the TABERNACLE. (Exodus 27:9-10) The same is said about the north side. (vs 11) Here are the TWENTY pillars with their TWENTY SILVER fillets and the TWENTY SILVER hooks, on each side of the court, south and north, the fillets and hooks being made of the SILVER that was given in REDEMPTION. Here again the number TWENTY and SILVER are connected with REDEMPTION.

The length of the court on the south and north is the same and their pillars, hooks and fillets are the same in number. This teaches that as many as receive the REDEMPTION of their souls will also receive the REDEMPTION of their bodies.

> "Whom He justified, them He also glorified." (Romans 8:29)

The glorified equal the justified. The number in one neither exceeds nor is less than the number in the other. Jesus said,

> "This is the will of him that sent me, that EVERY ONE which seeth the Son, and believeth on him, may have everlasting life: and I WILL RAISE HIM UP AT THE LAST DAY." (John 6:40)

This will glorify all who becomes justified. It will give the REDEMPTION of the body to EVERY ONE who receives the REDEMPTION of the soul. This is shown in the south and north sides of the Tabernacle with their TWENTY boards to the side and also in the sides of the court, south and north, equal in length, and in number of pillars. Let those who teach that a child of God can lose his salvation adjust these numbers and dimensions to his doctrine, if he can. IT CANNOT BE DONE!

In Ruth 4:1-10 there is the record of Boaz, the kinsman-redeemer, redeeming the property that was Elimelech's and Naomi's, and purchasing Ruth to be his wife. The name of Boaz occurs TWENTY times in the Book of Ruth. The names of TWENTY different persons are mentioned in the BOOK OF RUTH.

You will remember that Jacob labored in the household of Laban for TWENTY years in order to REDEEM Rachel. (Genesis 31:38-41)

In the wilderness all that had been numbered of the twelve tribes of Israel from TWENTY years old and upward were to die in the wilderness. Those under TWENTY years of age who, they feared would be a prey to their enemies, were to be the ones to go into the land. This shows that the fear of man is many times without foundation. When God is in a thing man never needs to fear the results. God is greater than all and will fulfill His Word. Those who walked by faith, experienced the REDEMPTION OF THE LORD. (Numbers 14:26-35)

TWENTY-ONE

EXCEEDING SINFULNESS OF SIN

The history of Israel's wilderness journey discloses that TWENTY-ONE sins were recorded against her from Egypt to Jordan. This number would seem to indicate THE EXCEEDING SINFULNESS OF SIN. I do not mean that TWENTY-ONE sins are all the sins one can commit, but God used that number to represent the EXCEEDING SINFULNESS OF SIN. A full understanding of these TWENTY-ONE sins can be reviewed by turning back to the study on number THIRTEEN, the number for DEPRAVITY. In that study it was seen that THIRTEEN multiplied by TWENTY-ONE equals 273, the number that needed to be redeemed with FIVE shekels per person. (Numbers 3:46-47) These two numbers, THIRTEEN and TWENTY-ONE are so very closely associated it would seem that TWENTY-ONE is the outgrowth of THIRTEEN, which represents the DEPRAVED NATURE, and that TWENTY-ONE is the FRUIT of that nature.

In II Timothy 3:2-5 Paul lists TWENTY-ONE things which men would do in the last days, and he warns against such.

> "In the last days perilous times shall come, for men shall be (1) lovers of their own selves, (2) covetous, (3) boasters, (4) proud, (5) blasphemers, (6) disobedient to parents, (7) unthankful, (8) unholy, (9) without natural affection, (10) truce breakers, (11) false accusers, (12) incontinent, (13) fierce, (14) despisers of those who are good, (15) traitors, (16) heady, (17) highminded, (18) lovers of pleasure, (19) more than lovers of God, (20) having a form of godliness, (21) but denying the power thereof: from such turn away."

In Matthew 23, we have recorded the TWENTY-ONE

characteristics of a hypocrite. Here is THE EXCEEDING SINFULNESS OF SIN and the Lord is condemning them.

1. Demand respect as teachers. (vs 2)
2. Teach, but do not practice. (vs 3)
3. Demand service; not give it. (vs 4)
4. Seek praise of men. (vs 5)
5. Parade their religion. (vs 5)
6. Seek chief banquet places. (vs 6)
7. Seek chief places in church. (vs 6)
8. Glory in personal attention. (vs 7)
9. Glory in titles. (vs 7)
10. Rob men of truth and life. (vs 13)
11. Reject truth and life. (vs 13)
12. Take advantage of widows. (vs 14)
13. Exhibit long prayers. (vs 14)
14. Are zealous to win men to their sect, but not to God. (vs 15)
15. Root and ground converts in hypocrisy, not in God. (vs 15)
16. Profess to be the only guide in religion, but are blind to truth. (vs 16-22)
17. Propagate those parts of religion from which they receive most personal gain and honor. (vs 16-22)
18. Strain at gnats and swallow camels; stress minor details and omit the fundamentals of religion. (vs 23-24)
19. Glory in bodily cleanliness, but live in moral filth. (vs 25-26)
20. Exhibit outward religion and self-righteousness and ignore inward holiness in life and conduct. (vs 2, 7-28)
21. Pretend to be more righteous than their forefathers. (vs 29-33)

This number is mentioned very few times in the Word of God but always reveals the EXCEEDING SINFULNESS OF SIN.

TWENTY-TWO
LIGHT

TWENTY TWO is the number that is connected with LIGHT. There were TWENTY-TWO bowls to hold oil in the candlestick in the Tabernacle. There were THREE branches on each side of the shaft of the candlestick. Each branch had three bowls. This makes six branches, with EIGHTEEN bowls. In the candlestick itself (the shaft) were FOUR bowls. (Exodus 25:31-34) This is a total of TWENTY-TWO bowls serving the SEVEN lamps. The purpose of the candlestick with its lamps was to give light.

In Matthew 5:15-16 Jesus said,

> "Neither do men light a candle and put it under a bushel, but on a candlestick and it giveth Light to all in the house. Let your LIGHT so shine before men that they may see your good works, and glorify your Father which is in heaven."

The saved are called the children of LIGHT. "Ye are all children of the LIGHT." (I Thess. 5:5) When FOURTEEN for salvation is added to EIGHT for the New Birth, the sum is TWENTY-TWO, LIGHT.

In Acts 22:4-11, Paul was relating his experience on the Damascus road, He told about the great light that shone from heaven. He said he heard a voice saying unto him, "Saul, Saul, why persecutest thou me?" In this place the name of Saul occurs the TWENTY-SECOND time.

In Numbers 3:39 a total of 22,000 Levites were numbered to serve in the Priestly work of the Tabernacle. They were to give and minister LIGHT to the people. The number THOUSAND means the "Glory of our Lord" and TWENTY-TWO is the number of LIGHT. Thus, the number 22,000 Levites were to reveal to the people the LIGHT of the GLORY

OF THE LORD. How can anyone deny the inspiration of the Scriptures?

The word LIGHT is found 264 times in the Bible which divided by TWELVE, the number of ADMINISTRATION, we have the number of LIGHT. In other words, we are to ADMINISTER the LIGHT of the World to the hearts of people. In the Gospel of John the word LIGHT is used TWENTY-TWO times. Amazing isn't it?

TWENTY-THREE
DEATH

TWENTY-THREE is the number that represents DEATH. This is found by reading Romans 1:28-32. There are TWENTY-THREE things listed that God said, "that they which commit such things are worthy of DEATH, not only do the same but have pleasure in them that do them."

This is in harmony with all that has been learned about the numbers previous to this one. TEN represents the LAW, and THIRTEEN the DEPRAVED, REBELLIOUS heart of man. SIN through the LAW brings DEATH, and TEN plus THIRTEEN equals TWENTY-THREE. Paul said,

> "I was alive without the law (10) once; but when the commandment (or law) came, sin (13) revived, and I died(23)". (Romans 7:9)

Paul said in I Corinthians 15:56,

> "The sting of DEATH (23) is sin and the strength of sin (13) is the law (10)."

It takes THIRTEEN plus TEN to equal TWENTY-THREE. So it takes SIN plus the LAW to bring DEATH.

This proves that DEPRAVITY alone does not bring spiritual death. Babies come into the world depraved. We are "By nature children of wrath." (Ephesians 2:3) So was Paul when he was a child. But THIRTEEN (sin) alone does not equal TWENTY-THREE (death). TEN more must be added. When the LAW (10) was added that made TWENTY-THREE (DEATH) and Paul died. This takes care of the charge brought against those who teach inherent DEPRAVITY. They are accused of teaching infant damnation. But this charge falls to the ground in the face of Bible numbers and the statement of Paul in Romans 7:9-11.

Now work an equation with THREE for the RESUR-

RECTION and TWENTY for REDEMPTION. When God's children are raised from the dead (Luke 20:35-36 and Revelation 20:4-6) they will be brought out from death. By subtracting THREE from TWENTY-THREE there are TWENTY left. This will give the saved the REDEMPTION of their bodies.

The numbers ELEVEN and TWELVE work in like manner. Only the constituted authority or judge can pronounce the judgment for any crime. ELEVEN for Judgment, plus TWELVE for DIVINE AUTHORITY equals TWENTY-THREE for DEATH. Read Romans 1:32 again,

> "Who knowing the judgment (11) of God (12 the Divine Judge), that they which commit such things are worthy of DEATH (23).

ELEVEN plus TWELVE equals TWENTY-THREE. See how the equation holds good?

Will the skeptic please tell why the application of the Bible numbers always work like this? He surely cannot say that a finite being like this writer could frame up such a complicated system of Bible numbers, and make them fit in all the verses so perfectly. It is too far reaching, and too frequent to be accidental. There can be only one answer. The infinite mind of God devised this system of numbers, and inspired men from Genesis to Revelation to so record them that they would all fit into their places. The same God who inspired men to write them in the Bible has enabled me, and others who have studied numbers, to discover them in His Word.

In studying the resurrection in I Corinthians 15th chapter, Paul used the word "RESURRECTION" four times, the word "RAISED" ten times, the word "RISEN" three times, the word "RISE" four times and the word "ROSE" twice. This is TWENTY-THREE in all and shows that the RESURRECTION WILL bring out of the state of DEATH, the believer, in a new body, which is represented by that number.

In Genesis 7:21-22 the record tells that all flesh died upon the earth. The next verse says, "Noah only remained alive, and they that were with him in the ark." This is the TWENTY-THIRD time the name of Noah is found.

In Genesis 19:24-25, there is the record of God raining fire and brimstone on Sodom and Gomorrah. In verse 27 and 28 it is said that Abraham got up early and looked toward Sodom and Gomorrah, and beheld the smoke of those cities going up. This is the TWENTY THIRD time the name of Abraham is found. The TWENTY-THIRD time the name of Jacob is found is where his mother tells him that his brother Esau purposes to kill him. (Genesis 27:42).

In Revelation 20:12 John said,

> "I saw the DEAD, small and great stand before God; and the books were OPENED."

This is the TWENTY-THIRD time the word "OPEN" is found in the book of Revelation.

In Revelation 17:1-4; 17:18; 18:6, John saw this woman (the harlot) sitting upon a beast with SEVEN heads and TEN horns. He saw her arrayed in SIX things. John was called to witness what would befall this woman. DEATH was to come to her. "Therefore shall her plagues come in one day, DEATH, and mourning, and famine." (Revelation 18:8)

When the above numbers, SEVEN, TEN, and SIX are added, the sum is TWENTY-THREE for DEATH.

In the same verse where it is said that DEATH had come upon this great city or woman, it is said,

> "Strong is the Lord God who JUDGETH her." (Revelation 18:8)

When ELEVEN for judgment is added to TWELVE for the Divine Authority, which passes judgment upon the whore, again the sum is TWENTY-THREE for DEATH.

Right after Peter's name occurs the TWENTY-THIRD time in the book of Acts, (Acts 9:34), Tabitha or Dorcas dies. (vs 36-37) Then Peter's name occurs THREE more times. (Vs 38, 39, and 40) And He raised Dorcas or Tabitha from the dead. So here is

TWENTY-THREE for death and THREE for the RESURRECTION. This makes the TWENTY-SIXTH time Peter's name occurs when he raises Tabitha from the dead. TWENTY-SIX is the number of the Gospel of Christ and the gospel is the good news about our Lord's death for our sins, and His resurrection after THREE days. (Read I Corinthians 15:1-4). TWENTY-THREE for His death, plus THREE for His resurrection make TWENTY-SIX for the Gospel. Those same numbers are found in connection with the death and resurrection of Dorcas. Her death follows the TWENTY-THIRD time Peter's name is found. When his name is mentioned THREE more times she is raised from the dead. That puts her resurrection the TWENTY-SIXTH time Peter's name is found in the book. So here is the number for GOOD NEWS, or the Gospel. It certainly was good news to the saints when Peter presented her to them alive. (vs 41). Marvelous is the Word of God!

TWENTY-FOUR

THE PRIESTHOOD

TWENTY-FOUR is the number associated with the PRIESTHOOD. If the reader will read I Chronicles 24:1-18, he will find David distributing the priesthood among TWENTY-FOUR of the descendants of Aaron. After Nadab and Abihu died and left no children, Aaron had two sons left, Eleazar and Ithamar. Among the sons of Eleazar there were SIXTEEN chief men, and of the sons of Ithamar there were EIGHT chief men. David made these to be governors of the sanctuary.

This number is carried over into the Book of Revelation. You will note the Scriptures in Revelation 4:4 which says,

> "And round the throne were FOUR and TWENTY seats: and upon the seats I saw FOUR and TWENTY elders sitting, clothed in white raiment; and they had on their heads crowns of gold." (Revelation 4:4)

Later on these TWENTY FOUR elders, together with the FOUR beasts (or living creatures) are singing that "Christ has redeemed them by His blood from every (1) kindred, (2) tongue, (3) people, and (4) nation; and has made them unto God kings and PRIESTS; and they shall reign on the earth." (Revelation 5:9-10)

A priest is one who intercedes for another, one who is a transgressor. There are TWENTY FOUR hours in a day and night. Sinful man needs a priest every hour of his life. Notice how Job continually made offerings for his children and prayed for them. (Job 1:5) While reckless carefree boys and girls are away from home, or while they are out at all hours, in places of danger, in places of sin, godly fathers and mothers are interceding for them day and night.

Notice that Eleazar had SIXTEEN sons and Ithamar had EIGHT. These numbers stand for LOVE and the NEW BIRTH. To rightly intercede for others one must have LOVE toward God, and toward the ones for whom he is interceding, and he must also be BORN AGAIN.

Christ is our High Priest. His people are eternally secure because He intercedes for them every hour (24) of the day and night. Listen to the Words of God,

> "Behold He that keepeth Israel shall neither slumber nor sleep . . . The sun shall not smite thee by day, nor the moon by night. The Lord shall preserve thee from all evil: He shall preserve thy soul." (Psalm 121:4-7)

In the SEVENTY-SECOND Psalm there are listed TWENTY-FOUR things that the Messiah will do for His people. Here He is the Intercessor upon His throne in heaven, who is going to sit upon His throne upon the earth and intercede for His people.

TWENTY-FIVE

THE FORGIVENESS OF SINS

TWENTY-FIVE is evidently the number for the FORGIVENESS OF SINS. Moses prayed for God to forgive the sins of Israel, and not to blot out that nation. He prayed,

> "Pardon, I beseech thee, the iniquity of this people according to the greatness of thy mercy, and as thou hast forgiven this people, from Egypt even until now." (Numbers 14:13-19)

Then God said to Moses, "I have pardoned according to my Word." (Vs 20) This forgiveness took place as the result of the INTERCESSION of Moses, and because of God's mercy, or GRACE. It followed the intercessory work of Moses. TWENTY-FOUR stands for the believer's priesthood. TWENTY-FIVE is the next number after TWENTY-FOUR. Forgiveness of sins follows intercession. Paul prayed for Israel that they might be saved.

> "Brethren, my heart's desire and prayer to God for Israel is, that they might be saved." (Romans 10:1)

The number TWENTY-FIVE is connected with a pardon for Jehoiachin, King of Judah. Read Jeremiah 52:31-32. Jeremiah states that Jehoiachin's prison garments were changed, and that he did eat bread before him all the days of his life. (vs 33) Here is a picture of a pardoned sinner. This happened on the TWENTY-FIFTH day of the month. What a beautiful picture of Christ who forgives our sins, changes our garments and feeds us on the bread of life all of our days.

When Ephesians 1:7 is studied in the light of numbers the same thought is brought out.

> "In whom we have redemption (20) through his blood, the FORGIVENESS OF SINS, according to the riches of His GRACE (5)." (Ephesians 1:7)

This verse contains REDEMPTION, FORGIVENESS OF SINS, and GRACE. TWENTY has been found to be the number for REDEMPTION, and FIVE the number of GRACE. When these two numbers are added it gives you TWENTY-FIVE for FORGIVENESS OF SINS.

The Levites were TWENTY-FIVE years old when they began their service at the TABERNACLE.

> "This is it that belongeth to the Levites: from TWENTY-FIVE years old and upward they shall go in to wait upon the service of the Tabernacle." (Numbers 8:24)

We are not ready to do service for God until our SINS HAVE BEEN FORGIVEN. From that time onward we can serve Him.

TWENTY-SIX
THE GOSPEL OF CHRIST

TWENTY-SIX seems to be the number that stands for the Gospel of Christ. Paul said to Timothy,

> "Be not thou therefore ashamed of the testimony of our Lord, nor of me his prisoner: but be thou partaker of the AFFLICTIONS OF THE GOSPEL according to the power of God." (II Timothy 1:8)

In II Corinthians 11:23-27 Paul listed TWENTY-SIX different afflictions which he had endured as a minister of Christ for the GOSPEL.

In I Corinthians 15:1-4 Paul states that the GOSPEL that he had declared was "That Christ died for our sins according to the Scriptures; and that He was buried, and that He arose again the third day according to the Scriptures." TWENTY-THREE is the number for DEATH, and THREE is the number for the RESURRECTION. These two numbers when added made TWENTY-SIX, the number of the GOSPEL OF CHRIST.

In John 3:16 there are exactly 26 words in the Greek Language. Outos (1) gar (2) agapasen (3) ho (4) Theos (5) ton (6) kosmos (7) hoste (8) tov (9) huion (10) autou (11) tov (12) monogena (13) edoken (14) hina (15) pas (16) ho (17) pisteuon (18) eis (19) auton (20) mn (21) apolatai (22) all (23) exa (24) zoan (25) aioviov (26). (John 3:16)

There are only 25 words in the English translation, but TWENTY-SIX in the Greek.

The New Testament was written in the Greek language. Number TWENTY-SIX stands for the GOSPEL, which means "good news". TWENTY-FIVE stands for FORGIVENESS OF SINS, which comes through believing in Jesus Christ of whom the GOSPEL speaks.

The TWENTY-SIXTH time the name of Noah occurs is in Genesis 8:11 where the dove returned to Noah with the olive leaf in her mouth. It goes on to say,

> "So Noah knew that the waters were abated from off the earth."

Here was good news for Noah, and the word GOSPEL means good news.

Right after Peter's name occurs the TWENTY-THIRD time in the book of Acts (Acts 9:34), Tabitha or Dorcas dies. (vs 36-37) Then Peter's name occurs THREE more times (vs 38, 39, 40) and he raised Tabitha from the dead. So here is TWENTY-THREE for death, and THREE for the RESURRECTION. This makes the TWENTY-SIXTH time Peter's name occurs when he raises Tabitha from the dead. TWENTY-SIX is the number for the GOSPEL, and the gospel is good news about our Lord's death for our sins, and His resurrection after THREE days. (I Corinthains 15:1-4) TWENTY-THREE for His death plus THREE for His resurrection make TWENTY-SIX for the GOSPEL. Those same numbers are found in connection with the death and resurrection of Dorcas or Tabitha. Her death follows the TWENTY-THIRD time Peter's name is found. When his name is mentioned THREE more times she is raised from the dead. That puts her resurrection the TWENTY-SIXTH time Peter's name is found in the book. It certainly was good news to the saints when Peter presented her to them alive. (41 vs.) This presentation follows the TWENTY-SEVENTH time Peter's name occurs, which suggest that TWENTY-SEVEN may stand for the preaching of the good news of the GOSPEL.

TWENTY-SEVEN

PREACHING OF THE GOSPEL

This number is found very few times in the Bible. The author has searched the Scriptures and it seems that this number is made up of a combination of numbers that reveals the PREACHING OF THE GOSPEL.

In I Timothy 3:1-7 there is listed SEVENTEEN qualifications for the Preacher of the Gospel. Number SEVENTEEN is the number for VICTORY. Number TEN is the number for the LAW. Together they give us the number TWENTY-SEVEN which attest to the fact that the Preacher is one who gives his testimony of VICTORY that the good news, the GOSPEL overcomes and frees from the LAW and gives VICTORY to the life of the believer.

TWENTY-EIGHT
ETERNAL LIFE

Since we have seen that TWENTY-SIX is the GOSPEL OF CHRIST and TWENTY-SEVEN is the PREACHing OF THE GOSPEL, then TWENTY-EIGHT is the number for ETERNAL LIFE. ETERNAL LIFE always is the results for the PREACHING OF THE GOSPEL OF CHRIST. This number is found by combining some numbers that are found in certain passages of Scripture where the words ETERNAL LIFE are found. This is also true in the previous number, TWENTY-SEVEN.

In Romans 5:20-21 Paul said,

> "But where sin abounded, GRACE did much more abound: that as sin hath reigned unto DEATH (23), even so might GRACE (5) reign through righteousness unto ETERNAL LIFE by Jesus Christ our Lord."

These two numbers add up to ETERNAL LIFE, and the expression ETERNAL LIFE is in this passage.

Romans 5:15 speaks of "The Gift by GRACE". Romans 6:23 says, "The wages of sin is DEATH (23); but the gift of God is ETERNAL LIFE through Jesus Christ our Lord." By adding 23 for DEATH and FIVE for GRACE, the sum obtained is TWENTY-EIGHT, which stands for ETERNAL LIFE.

The same thing is seen in John 5:24. In this verse Jesus makes FIVE (5) positive statements. It has been proven that FIVE stands for GRACE. In the FIFTH of these statements Jesus said, "But have passed from DEATH (23) unto LIFE (28)." An examination of this verse follows:

1. "He that heareth my word,
2. And believeth on Him that sent me,
3. Hath everlasting life,
4. And shall not come into condemnation;
5. But is passed from DEATH unto LIFE."

In the FIVE divisions of the verse, GRACE is displayed. In the FIFTH division man has "Passed from DEATH unto LIFE." The life under consideration is everlasting, or ETERNAL LIFE. It is God's GRACE by which one passes from DEATH UNTO LIFE. By adding FIVE for GRACE to TWENTY-THREE for DEATH the sum is TWENTY-EIGHT, the number for ETERNAL LIFE.

Another combination of numbers that make TWENTY-EIGHT, when added, is EIGHT for the NEW BIRTH, and TWENTY for REDEMPTION. Hebrew 9:12 states that Christ obtained ETERNAL REDEMPTION for us. Certainly in the ETERNAL REDEMPTION, a child of God has eternal salvation. In the NEW BIRTH he receives the very life of God, which is ETERNAL.

The linen curtains in the Tabernacle were TWENTY-EIGHT cubits long.

> "The length of one curtain shall be eight and twenty cubits." (Exodus 26:1-2)

> "The FIVE curtains shall be coupled together, one to another." (Verse 3)

This connects GRACE with the curtains TWENTY-EIGHT cubits long. And GRACE is connected with ETERNAL LIFE.

In Romans 5:21 Paul states,

> "That as sin hath reigned unto death, even so might GRACE reign through righteousness unto ETERNAL LIFE by Jesus Christ our Lord."

Notice that DEATH is (23), GRACE is (5), and together form 28, ETERNAL LIFE. It is through the GRACE OF GOD that DEATH is conquered and ETERNAL LIFE is secured.

The TWENTY-EIGHTH time Noah's name is found is in Genesis 8:15,

> "And God spake unto Noah, saying, Go

> forth of the ark, thou, and thy wife, and thy sons, and thy sons' wives with thee." (Genesis 8:15-16)

In Noah and his family going into the new age there is a picture of those who have ETERNAL LIFE entering into the bliss of a new age. Truly the Word of God is marvelous.

This writer has just checked upon the number of Greek words in John 3:14-15, and he finds that there are exactly TWENTY-EIGHT, and the last two are the words for ETERNAL LIFE. While the English has more than TWENTY-EIGHT words, yet there are only TWENTY-EIGHT in the original language. These TWENTY-EIGHT words in John 3:14-15, together with the TWENTY-SIX words in John 3:16, as has been shown, make exactly FIFTY-FOUR words, the exact number of pillars in the court of the TABERNACLE.

NOTE: This writer is trying to check the number of words in the original language, as well as the English translation. Where the numbers are the same as in the original and the translation only the English words are given. Otherwise the Greek is given. If the original words do not count out the number is not used.

TWENTY-NINE
DEPARTURE

Number TWENTY-NINE is a number that is associated with DEPARTURE, or GOING AWAY. The TWENTY-NINTH time the name of Noah is found is in the place where he, and all that were with him, went forth out of the ark. (Genesis 8:18-19)

The TWENTY-NINTH time the name Abram occurs is where the kings who defeated the kings of Sodom and Gomorrah took Lot and all his goods and DEPARTED. (Genesis 14:12)

The TWENTY-NINTH time the name of Abraham (not Abram) occurs is where he tells Abimelech the agreement he had with Sarah when he left his Father's house. (Genseis 20:11-13)

The TWENTY-NINTH time the name of Isaac occurs is where he went unto Abimelech in Gerar. (Genesis 26:1) Before this he had dwelt by the well, Lahairoi. (Genesis 25:11)

The TWENTY-NINTH time the name of Jacob is found is in the place where Jacob had left Canaan and had gone to Padan-aram. (Genesis 28:7)

The TWENTY-NINTH time the name of Laban's is found is in the place where God told Jacob to leave Laban's place and to return to the land of his father's.

The TWENTY-NINTH time the name of Samson occurs is where he awoke out of his sleep and WENT AWAY with the pin of the beam and the web. (Judges 16:14)

This number is found very few times in the Bible.

It would be well for the reader to compare the TWO numbers, TWENTY AND NINE. TWENTY stands for REDEMPTION and NINE for the Fruit of the Spirit. Together they reveal that the Christian who has been redeemed should bear FRUIT for the Master, the Fruit of the Spirit.

THIRTY
THE BLOOD OF CHRIST - - - DEDICATION

THIRTY is the number for the BLOOD OF CHRIST. Judas Iscariot betrayed Jesus for THIRTY pieces of silver. (Matthew 26:14-15) After having betrayed Jesus for THIRTY pieces of silver, Judas repented himself and brought the money back. The chief priests took the silver pieces and said,

> "It is not lawful for to put them in the treasury, because it is THE PRICE OF BLOOD."

Then they took counsel and bought with them the potter's field, to bury strangers in. And the Bible says, "Wherefore that field was called, the field of BLOOD, unto this day." (Matthew 27:3-8)

This teaches that THIRTY is the number for THE BLOOD OF CHRIST. REDEMPTION is through the BLOOD of Christ.

> "In whom we have REDEMPTION through His BLOOD." (Ephesians 1:7)

Christ came to redeem men from the law.

> "God sent forth His Son, made of a woman, made under the law, to REDEEM them that were under the LAW." (Galatians 4:4-5)

By adding TWENTY-for REDEMPTION to TEN for the LAW the sum is THIRTY, the price of REDEMPTION from the LAW, which was the BLOOD of CHRIST.

Now add TWENTY-FIVE for FORGIVENESS OF SINS to FIVE for GRACE and the answer is THIRTY.

> "In whom we have REDEMPTION through His BLOOD, the FORGIVENESS of sins, according to the riches of HIS GRACE." (Ephesians 1:7)

The THIRTIETH time the name of Noah is found is in Genesis 8:20 where he built an altar unto the Lord and offered of every clean beast unto the Lord. In this connection it is said, that God smelled a sweet savour, and He promised not to curse the ground any more for man's sake. Paul said that Christ gave "Himself for us an offering and a sacrifice unto God for a sweet smelling savour." (Ephesians 5:2) So the THIRTIETH time Noah's name is mentioned there is a picture of the BLOOD OF CHRIST being shed.

The number THIRTY is connected with the LIFE of CHRIST. Jesus began His public ministry at the age of THIRTY. (Luke 3:23) According to the law of God, the Priests could not serve in his capacity until he was THIRTY years of age. This was the AGE OF DEDICATION for offering the BLOOD upon the altar. (Numbers 4:3; 1 Chronicles 23:2-3) Jesus is our High Priest and to fulfill the plan of His Father, He would not, and could not, begin His public ministry under THIRTY years of age, thus, fulfilling the requirements of PRIESTHOOD.

John the Baptist was of the priestly tribe. His father was a Priest. (Luke 1:5) The first born son was to be a priest. John was the first born. John the Baptist was a priest and was six months older than Jesus. (Luke 1:26) He began his public ministry six months before Jesus after reaching the age of THIRTY. As our High Priest, Christ is to offer the BLOOD upon the altar before God for the atonement for sins. John prepared the way for the High Priest.

The goat's hair curtains on the Tabernacle were THIRTY cubits long. (Exodus 26:7-8) The ELEVEN curtains of goat's hair represent Jesus bearing the judgment for our sins, since ELEVEN is the number for JUDGMENT. He did this when His BLOOD was shed on the cross. The BLOOD is represented by the THIRTY cubits, the length of the curtains.

FIVE of the goat's hair curtains were coupled together, and the other SIX were coupled together. (Exodus 26:9) These two broad ones were fastened together with FIFTY taches of brass. (vs 10-11) The sixth curtain was doubled over the forefront of the Tabernacle. This made the brass taches in the goat's hair curtains, as well as the gold taches in the linen curtains, to come over the veil. "Thou shalt hang up the veil under the taches." (Exodus 26:33) See also verse 9. This veil represented the flesh of Jesus that was rent on the cross. (Hebrews 10:19-20) Thus in the veil is displayed a picture of the cross. The BLOOD of the cross is portrayed in the length of the goat's hair curtains, THIRTY cubits, meeting at the veil. Judgment being met by the BLOOD of Christ, which was shed on the cross, is shown by the number of goat's hair curtains, ELEVEN. What wondrous wisdom is displayed in all this! Was there ever such a book as the Bible? With what surprising wisdom did God plan all these things! THIRTY is mentioned 168 times in the Bible.

THIRTY-ONE
OFFSPRING

THIRTY-ONE is a number that is connected with OFFSPRING. The THIRTY-FIRST time Noah's name occurs is where God said, to him and his sons,

> "Be fruitful and multiply, and replenish the earth." (Genesis 9:1)

The THIRTY-FIRST time the name of Abram occurs is where it speaks of his trained servants BORN in his house. (Genesis 14:14)

The THIRTY-FIRST time the name Abraham occurs is where he prayed for the house of Abimelech, and the wife and the maidservants of Abimelech bare children. (Genesis 20:17-18)

The THIRTY-FIRST time the name of Jacob is found is when he awaked out of his sleep after God had told him that His SEED would be as the dust of the earth. (Genesis 28:13-16)

THIRTY-ONE is the number after THIRTY, which it seems stands not only for the BLOOD of Christ, but BLOOD in any case. The THIRTIETH time Noah's name occurs is where he offered sacrifices unto the Lord. (Genesis 8:20) It has already been shown that THIRTY is connected with the BLOOD of Christ in three places. (Matthew 27:3-8)

The THIRTIETH time Samuel's name occurs is where he offered a lamb for a burnt offering. (I Samuel 7:9)

In the same connection where Paul said that God had made of one BLOOD all nations of men he spoke about men being the OFFSPRING of God. The statement about the OFFSPRING follows the statement about the BLOOD, even as THIRTY-ONE follows right after THIRTY. (Acts 7:26-29)

In Proverbs 31:10-31, there are listed THIRTY-ONE virtues of a model woman. In the 28 and 29 verses her OFFSPRINGS will arise up and call her blessed. This number is found only sixteen times in the Bible.

THIRTY-TWO
COVENANT

The number THIRTY-TWO is associated with a COVENANT. The THIRTY-SECOND time Noah's name is found is where God made a COVENANT with him. (Genesis 9:8-9)

> "And God spake unto Noah, and to his sons with him, saying, And I, behold, I establish my covenant with you, and with your seed after you:"

There are THIRTY-TWO references in the book of Deuteronomy to the Abrahamic COVENANT.

The COVENANT was made with Noah and his seed (OFFSPRING) right after God gave to Noah and his sons the command to be fruitful and multiply. (Genesis 9:1-9)

The name of Boaz, who redeemed the property of Naomi and Elimelech, (Ruth 4:1-10) is found TWENTY times. The name of Ruth is found TWELVE times in the same book. These two numbers when added make THIRTY-TWO, and Christ, through whom God's COVENANT to Abraham is fulfilled, was a descendant of Boaz and Ruth. This number is found THIRTY-TWO times in the Bible.

THIRTY-THREE
PROMISE

THIRTY-THREE is the number that is associated with PROMISE. The THIRTY-THIRD time Noah's name is found is where God gave the rainbow as a token of His promise to never again destroy all flesh with a flood. (Genesis 9:13-17)

Isaac was a child of PROMISE.

> "Now we, brethren, as Isaac was, are the children of PROMISE." (Galatians 4:28)

Abraham's name is found the THIRTY-THIRD time in the place where Isaac, the child of PROMISE is born. In the same place the record speaks of God visiting Sarah as He had spoken, or PROMISED. (Genesis 21:1-2)

The THIRTY-THIRD time the name of Jacob is found is where he PROMISED to give God a tenth of all God gave to him. (Genesis 28:20-22)

Jesus was THIRTY-THREE years of age when He was crucified on the Cross and arose from the grave. Through His death, burial and, resurrection we have the PROMISE to live with Him for eternity. THIRTY is the number for THE BLOOD OF CHRIST and THREE is the number for RESURRECTION. Through the BLOOD OF CHRIST we have the PROMISE of a glorious RESURRECTION.

THIRTY-FOUR
NAMING OF A SON

This number is mentioned only once in the Bible. This is found in Genesis 11:16,

> "And Eber lived four and thirty years, and begat Peleg."

The THIRTY-FOURTH time Abraham's name is mentioned is in Genesis 21:3,

> "And Abraham called the name of his son that was born unto him, whom Sarah bare to him, Isaac."

This number seems to have a meaning of NAMING OF A SON.

THIRTY-FIVE
HOPE

THIRTY-FIVE is the number connected with HOPE. This number is used very seldom in the Word of God. But other numbers placed together reveals the meaning of this number. This is revealed in I Corinthians 13:13,

"And now abideth faith, HOPE, charity."

The number for faith is NINETEEN, and the number for charity, or love, is SIXTEEN, and these two numbers make THIRTY-FIVE. Should the critic object to adding the numerical value of words in this manner then let us consider how they work in Ephesians 2:8.

"By GRACE are ye saved through faith."

The number for GRACE has been shown to be FIVE, and the number for salvation is FOURTEEN, and these two numbers when added make NINETEEN, the number for faith. In II Thessalonians 2:18, Paul states that good HOPE comes through GRACE. In Ephesians 1:7 he shows that grace comes through the blood of Christ. When FIVE for GRACE is added to THIRTY for the BLOOD OF CHRIST, the resulting sum is THIRTY-FIVE, the number for HOPE.

Hebrew 4:9 states, "There remaineth therefore a rest for the children of God." That rest will be entered when our bodies are redeemed. In Romans 8:23-24, the redemption of our body is connected with HOPE. When FIFTEEN for REST is added to TWENTY for redemption, the sum is THIRTY-FIVE for HOPE.

THIRTY-SIX
ENEMY

The number THIRTY-SIX stands for ENEMY. After Esther had told King Ahasuerus of the plot to destroy her people the record goes on to state,

> "Then king Ahasuerus answered and said unto Esther the queen, who is he, and where is he, that durst presume in his heart to do so." (Esther 7:5)

This is the THIRTY-SIXTH time the name of Esther is found. Then she said,

> "The adversary and ENEMY is this wicked Haman. Then was Haman afraid before the king and queen." (vs. 6)

In this place the name of Haman occurs the THIRTY-SIXTH time.

The SIXTH time Haman's name occurs is where he purposed to destroy all the Jews. (Esther 3:6) This was a violation of the SIXTH commandment which says, "Thou shalt not kill." (Exodus 20:13)

The Devil's number is SIX, and Jesus called him a murderer. (John 8:44) If we would multiply these two numbers we would get THIRTY-SIX, the ENEMY of God's cause.

The THIRTY-SIXTH time David's name is found is in 1st Samuel where he overcomes Goliath, his ENEMY. (I Samuel 17:50) The number SIX is found connected with this giant twice. The number SIX is found in his height. (vs. 4) He had SIX pieces of armour. (vs. 5-7) SIX times SIX is THIRTY-SIX. David went out to meet him with his faith fixed in God, and gained the VICTORY. (I Samuel 17:37-50) When NINETEEN for faith is subtracted from THIRTY-SIX for ENEMY the remainder is SEVENTEEN, exactly the number for VICTORY. Will the modernist and infidel tell us why it works out this way?

Death is an ENEMY. "The last ENEMY that shall be destroyed is DEATH." (I Corinthians 15:26) "Sin entered the world and DEATH by SIN." (Romans 5:12)

When TWENTY-THREE for DEATH is added to THIRTEEN for SIN, the sum is THIRTY-SIX for ENEMY. The believer has the promise of a resurrection which will take him out of DEATH, the ENEMY. When THIRTY-THREE for promise is subtracted from THIRTY-SIX for ENEMY, the remainder is THREE, the number for the RESURRECTION. Now we are enabled to see the marvelous wisdom of God in arranging His system of numbers.

The THIRTY-SIXTH time the name of Abram occurs is where God said,

> "Fear not Abram: I am thy shield." (Genesis 15:1)

A shield is for protection from an ENEMY. None but God could so arrange these things.

In Revelation 12:7-11 the Dragon is pictured as an ENEMY. He fights against Michael and his angels. He accuses the brethren. John said,

> "They OVERCAME (victory) him by the blood of the Lamb, and by the word of their testimony; and they loved not their lives unto the DEATH."

When TWENTY-THREE for DEATH is added to THIRTY for BLOOD of Christ, the sum is FIFTY THREE. When THIRTY-SIX for the ENEMY (the one overcome) is subtracted from FIFTY-THREE, the remainder is SEVENTEEN for VICTORY.

Note the Scripture in I Corinthians 15:55-57 which says,

> "O death, where is thy sting? O grave, where is thy victory? The sting of death is SIN; (13) and the strength of sin is the LAW

> (10) But thanks be to God, which giveth us the VICTORY (17) through our Lord Jesus Christ''.

SIN (13) plus the LAW (10) brings Death (23). Through our FAITH in CHRIST (19) we overcome the ENEMY (36) and VICTORY is assured us for time and eternity. NINETEEN subtracted from THIRTY-SIX leaves SEVENTEEN the number for VICTORY.

This number is found TWENTY-THREE times in the Bible.

THIRTY-SEVEN
THE WORD OF GOD

Of all the numbers, this is perhaps the most sublime, setting forth the wonders of the written Word, and revealing through its pages Him, Who is the living Word of God!

THIRTY SEVEN occurs first in the numerical value of the two Hebrew words, "BRAYSHITH ELOHIM" -- "In the beginning God". The number of this phrase is 999, or THIRTY-SEVEN times TWENTY-SEVEN. This is a remarkable combination, for TWENTY-SEVEN is the cube of THREE. That number, THREE, means DIVINE COMPLETENESS or RESURRECTION, and cubing it, or taking it to its third power, tells of DIVINE PERFECTION made a reality, given substantial meaning. This humber TWENTY-SEVEN reveals the PREACHING OF THE GOSPEL OF GRACE and the word "KAPPORETH" -- "mercy seat," - occurs exactly TWENTY-SEVEN times in the Old Testament.

So here in the first two words of the Bible we have an arithmetical expression of the reality of the PREACHING OF THE GOSPEL OF CHRIST and of the Word of God (37), that Word, Who was in the beginning, Who was with God, and Who was God, as we learn in John 1:1.

Let us notice this number THIRTY-SEVEN. Jesus Christ is the very Word of God, the promised Saviour. Before He is born Joseph is told by an angel that he is to

> "call His name Jesus, for He shall save His people from their sins." (Matthew 1:21)

Now the number of this Name, Jesus, which is in Greek, Iesous, is 888, the numerical value, and the factors of this number are THIRTY-SEVEN and TWENTY-FOUR. Now THIRTY-SEVEN signifies the Word of God, and TWENTY-FOUR is the number of the PRIESTHOOD. Hence we see in the arithmetic of this sacred Name the same fact stated, that Christ is our High Priest who intercedes before the Father for us. (Hebrews 7:25-26 and Romans 8:34).

In the Old Testament we have God's Covenant with Abraham, given in Genesis 17, the Covenant of Circumcision, whereby every Jewish male child that was circumcised on the eighth day of its life, entered into covenant relationship with God, and became one of God's people. While the man child that was uncircumcised was to be cut off, "he hath broken My covenant." These were God's two words to the earthly seed of Abraham, corresponding to the two "whosoever" of the Gospel Message. The Holy Spirit emphasizes this Covenant, by using the Hebrew word "mul" -- "circumcise", exactly THIRTY-SEVEN times in the Old Testament, and the Hebrew word "arel" -- uncircumcised, exactly THIRTY-SEVEN times as well! Could there be a clearer proof that One Divine Mind inspired the whole Old Testament!

Further, the Old Testament points forward to the day when the Promised Seed of the woman should be born, and foretell in Micah 5:2,

> "But thou, Bethlehem Ephratah, though thou be little among the thousands of Judah, yet out of thee shall he come forth unto me that is to be ruler in Israel; whose goings forth have been from of old, from everlasting."

that He should be born in Bethlehem of Judah (there was another Bethlehem in Zebulun). Now Bethlehem in Judah, is mentioned exactly THIRTY SEVEN times in the Old Testament, because it was God's Word that the LIVING WORD should be born there! When He comes and has been born there, the name Bethlehem then occurs EIGHT times only in the New Testament, because the Word has been fulfilled, and the Seed has come Who is to give the NEW BIRTH, THE NEW BEGINNING, AND TO MAKE A NEW CREATION.

Again, the Golden Candlesticks of the Tabernacle and the Temple are undoubtedly a type of Him Who is the Light of the World, and these Candlesticks are mentioned exactly THIRTY-SEVEN times!

There is one outstanding character in the Bible of whom it is said, and that SIX times over, that he "wholly followed the Lord". This striking man was Caleb, one of the 12 spies sent to spy out the land of Canaan, and one of the two who brought back a good report. He knew that the Lord was with them, and that He would fight for them against their enemies. Caleb was thus the great Old Testament exponent of the overcoming life through faith in a present Saviour, and his name occurs just THIRTY-SEVEN times in the Old Testament.

Many very significant words are used by the Holy Spirit exactly THIRTY-SEVEN times, or a multiple of THIRTY-SEVEN, both in the Old and the New Testaments. We have only room to note a few of the most interesting.

PROSEUCHE -- "PRAYER" - is used in the New Testament exactly THIRTY-SEVEN times, and we can easily see why. Because prayer is only effectual when made in the Name of Him Who is the Word of God.

"No man cometh unto the Father but by Me." (John 14:6)

GAAL - - - - - "REDEEM" - is used 111 times (37 x 3) in the Old Testament, thus indicating that the Word of of God was the One coming to effect that wonderful redemption of sinners.

LAQAT - - - - "TO GLEAN" - occurs THIRTY-SEVEN times in the Old Testament, thus emphasizing the work of these who have enlisted under the Lord of the Harvest, and give heed to His last command before His Ascension.

The number THIRTY-SEVEN has a very remarkable mathematical property, in that it unifies the product whenever it is multiplied by THREE or any multiple of THREE up to its THIRD power. Thus 3 times 37 has the product of 111; 6 times 37 has the product of 222; 9 times 37 has the product of 333! It is the ONLY number that has this faculty, and hence its peculiar appropriatness to typify the Word of God, for the written Word is the only Book that can bring real unity of hearts to the different classes and races of mankind. It overleaps all barriers, and brings into closest bonds of fellowship by leading all to the Feet of the Living Word, the Lord Jesus Christ, so that they are "all one in Christ Jesus."

THIRTY-EIGHT

SLAVERY

This number is one of the few numbers in the Bible that is not mentioned very many times. It is found THREE times in the Old Testament and only once in the New Testament. Each time it is mentioned the thought of SLAVERY is revealed. In the Old Testament the people are enslaved by their own unbelief; by their king and by a nation. In the New Testament we see a man who is enslaved by disease.

The first time this number is used is in Deuteronomy 2:14,

> "And the space in which we came from Kadesh-barnea, until we were come over the brook Zered, was THIRTY and EIGHT years; until all the generation of the men of war were wasted out from among the host, as the Lord sware unto them."

Because of unbelief in the camp of God's people, judgment fell upon the Israelites. They had become slaves to their fears instead of walking by faith as a child of God.

The second time THIRTY-EIGHT is mentioned is in I Kings 16:29,

> "And in the THIRTY and EIGHTH year of Asa king of Judah began Ahab the son of Omri to reign over Israel: and Ahab the son of Omri reigned over Israel in Samaria twenty and two years."

It was during the reign of Ahab that the people of God were placed in slavery and idol worship became preeminent among the people.

The third time THIRTY-EIGHT is mentioned is in II Kings 15:8,

> "In the THIRTY and EIGHTH year of Azariah

> king of Judah did Zachariah the son of Jeroboam reign over Israel in Samaria six months."

It was during the reign of Zachariah that evil was done in the sight of the Lord and the people were ENSLAVED by the sins of the King and walked in the ways of Jeroboam the son of Nebat, who made Israel to sin.

The last time THIRTY-EIGHT is mentioned is in John 5:5,

> "And a certain man was there which had an infirmity THIRTY and EIGHT years."

Here is a condition of a man who had been afflicted for THIRTY-EIGHT years, the very length of time that the children of Israel wandered in the wilderness to learn the lesson of their sin and unbelief. This man is in utter helplessness. Nothing had availed for this paralytic. He could never reach the pool. He is enslaved with a disease of the body. Only Jesus could help him. Outside strength was needed. Only Jesus has the power to free from slavery whether of the body or soul. So sinners are saved through faith in the Word of God today.

THIRTY-NINE
DISEASE

The number THIRTY-NINE is found only once in the Word of God. This is recorded in II Chronicles 16:12,

> "And Asa in the THIRTY and NINTH year of his reign was diseased in his feet, until his disease was exceeding great: yet in his disease he sought not to the Lord, but to the physicians."

This number is connected then with DISEASE.

FORTY

TRIALS, PROBATION, AND TESTINGS

This number has long been universally recognized as an important number, both on the account of the frequency of its occurrence, and the uniformity of its association with a period of TRIAL, PROBATION and TESTINGS. It is the product of FIVE and EIGHT, and points to the action of GRACE (5), leading to and ending in REVIVAL and NEW BEGINNING OR NEW CREATION (8). This is certainly the case where FORTY relates to a period of evident PROBATION. But where it relates to ENLARGE DOMINION, or to RENEWED or EXTENDED rule, then it does so in virtue of its factors FOUR and TEN, and in harmony with their signification.

This number is mentioned 146 times in the Bible.

There are FIFTEEN such periods of Probation which appear on the surface of the Scriptures, and which may be thus classified:

1. FORTY YEARS OF PROBATION BY TRIAL:
   Israel in the Wilderness. (Deut. 8:2-5; Psalm 95:10; Acts 13:18)
   Israel from the crucifixion to the destruction of Jerusalem.
2. FORTY YEARS OF PROBATION BY PROSPERITY IN DELIVERANCE AND REST:
   Under Othniel. (Judges 3:11)
   Under Barak. (Judges 5:31)
   Under Gideon. (Judges 8:28)
3. FORTY YEARS OF PROBATION BY PROSPERITY IN ENLARGED DOMINION:
   Under David. (II Samuel 5:4)
   Under Solomon. (I Kings 11:42)
   Under Jeroboam II. (II Kings 12:17-18; 13:3, 5, 22, 25)
   Under Jehoash. (II Kings 12:1)
   Under Joash. (II Chronicles 24:1)
4. FORTY YEARS PROBATION BY HUMILIATION AND SERVITUDE:

Israel under the Philistines. (Judges 13:1)
Israel in the time of Eli. (I Samuel 4:18)
Israel under Saul. (Acts 13:21)

5. FORTY YEARS PROBATION BY WAITING:
Moses in Egypt. (Acts 7:23)
Moses in Midian. (Acts 7:30)

There are EIGHT great periods of testings revealed in the Word of God.

1. Moses was in the mountain of Sinai FORTY days and nights receiving the LAW. (Exodus 24:18) While he was gone these FORTY days the people became impatient and said to Aaron,

"Up, make us gods, which shall go before us; for as for this Moses, the man that brought us out of the land of Egypt, we know not what has become of him." (Exodus 32:1)

2. This led to the making of the golden calf of Exodus 32:2-7. Thus, Israel fell under this FORTY days of testing. (Deuteronomy 9:18,25)
3. After this they were tried FORTY years in the wilderness. (Numbers 14:34)

"After the number of the days in which ye searched the land, even FORTY days, each day a year, shall ye bear your iniquities, even FORTY years, and ye shall know my breach of promise."

4. FORTY days Elijah spent in Horeb after his experience on Mt. Carmel.

"And he arose, and did eat and drink, and went in the strength of that meat FORTY days and FORTY nights unto Horeb the mount of God." (I Kings 19:8)

5. FORTY days Jonah preached judgment would come to the city of Ninevah.

"And Jonah began to enter into the city a day's journey, and he cried, and said, Yet FORTY days, and Nineveh shall be overthrown." (Jonah 3:4)

6. FORTY days Ezekiel laid on his right side to symbolize the FORTY years of Judah's transgression.

"And when thou hast accomplished them, lie again on thy right side, and thou shalt bear the iniquity of the house of Judah FORTY days I have appointed thee each day for a year." (Ezekiel 4:6)

7. Our Saviour was tempted FORTY days and nights of the Devil.

"And Jesus, being full of the Holy Ghost returned from Jordan, and was led by the Spirit into the wilderness, being FORTY days tempted of the Devil." (Luke 4:1-2)

8. FORTY days Jesus was seen of His disciples, speaking of the things pertaining to the Kingdom of God.

"To whom also he shewed himself alive after his passion by many infallible proofs, being seen of them FORTY days, and speaking of the things pertaining to the Kingdom of God." (Acts 1:3)

The natural man, represented by number FOUR, falls under temptation. Romans 8:3 tells us how the LAW (10) was weak through the flesh (4):

"For what the law could not do in that it was weak through the flesh. God sending His own Son in the likeness of sinful flesh, and for sin, condemned sin in the flesh: that the righteousness of the law might be fulfilled in us, who walk not after the flesh, but after the Spirit."

Number FOUR which represents the natural man, multiplied by TEN, which represents the LAW, equals FORTY, and shows the man in the flesh falling under temptation and trials.

EIGHT, the number for the NEW BIRTH, multiplied by FIVE, the number for GRACE, equals FORTY, and shows the child of God standing up under temptation and testings.

> "God is faithful who will not suffer you to be tempted above that ye are able to bear it." (I Corinthians 10:13)

TWENTY, REDEMPTION for the soul, added to TWENTY, REDEMPTION for the body, makes FORTY, and put those who are REDEEMED beyond the reach of trials and temptations.

Moses spent FORTY years in Egypt; FORTY years in the desert; FORTY years with Israel in the wilderness. These years were times of trials, testings and temptations.

The TWELVE spies were commissioned to go and search out the land of Canaan to see if it was good or bad; fat or lean; wooded or barren. (Numbers 13:18-20) The Bible says,

> "And they returned from searching of the land after FORTY days." (Numbers 13:25)

The reigns of Saul, David, and Solomon each lasted FORTY years. Goliath defied Israel for FORTY days. Nineveh was given FORTY days to repent and turn back to God. (Jonah 3:4) Elijah fastened FORTY days and FORTY nights. Jesus was tempted FORTY days and FORTY nights and appeared ELEVEN times during FORTY days after His Resurrection. Punishment by flogging was limited to FORTY stripes save one. All these instances show that God was not hasty in His judgments, but gave man ample time for a fair trial.

FORTY-TWO

ISRAEL'S OPPRESSION - - LORD'S ADVENT

FORTY-TWO is the number that is associated with ISRAEL'S OPPRESSION, and the LORD'S ADVENT to the earth; both HIS first and second coming.

There were FORTY-TWO generations from Abraham to the FIRST ADVENT of Christ.

> "So all the generations from Abraham to David are fourteen generations; and from David until the carrying away into Babylon are fourteen generations; and from the carrying away into Babylon unto Christ are fourteen generations." (Matthew 1:17)

Three times FOURTEEN generations are FORTY-TWO generations. Therefore, Christ came into the world the first time FORTY-TWO generations from Abraham. This connects His first advent with the number FORTY-TWO. Is this without significance? Certainly God arranged it that way. It was His plan and purpose that it would be just FORTY-TWO generations from Abraham until the promised seed, Christ, who is the seed of Abraham. (Galatians 3:16)

> "Not to Abraham and his seed were the promises made. He said not, And to seeds, as of many; but as of one, And to thy seed, which is Christ."

should come the first time.

> "When the fullness of time was come, God sent forth His Son, made of a woman." (Galatians 4:4)

In the quotation from Matthew 1:17 both the word "David" and "Babylon" are found twice. TWO is the number for DIVISION; and those FORTY-TWO generations are divided into three periods of FOURTEEN generations each. THREE is the number for RESURRECTION, AND FOURTEEN is the number for DELIVERANCE. How well this fits with the passage in Hebrews 2:14-15,

> "Forasmuch then as the children are partakers of flesh and blood, he also himself likewise took part of the same; that through death he might destroy him that had the power of death, that is, the devil; and DELIVER them who through fear of death were all their lifetime subject to bondage."

It is through death and the resurrection that Christ brings this DELIVERANCE. THREE for the resurrection multiplied by FOURTEEN for deliverance makes FORTY-TWO, the number of the fullness of time spoken in Galatians 4:4.

Our Lord SECOND ADVENT to the earth will also be associated with the number FORTY-TWO. At the end of the FORTY-TWO months of Israel's oppression by the beast, Christ will make His SECOND ADVENT to the earth.

> "Then shall that wicked be revealed, whom the Lord shall consume with the spirit of His mouth, and shall destroy with the brightness of His coming." (II Thess. 2:8)

> "And power was given unto him (the beast) to continue FORTY and TWO months." (Revelation 13:5)

> "But the court that is without the temple leave out and measure it not; for it is given unto the Gentiles; and the holy city shall they tread under foot FORTY and TWO months." (Revelation 11:2)

> "And to the woman were given two wings of a great eagle, that she might fly into the

> wilderness, into her place, where she is nourished for a time, and times, and half a time, from the face of the serpent." (Revelation 12:40

> "It shall be for a time, times, and a half; and when he shall have accomplished to scatter the power of the holy people; all these things shall be finished." (Daniel 12:7)

The time, times, and half a time the woman (remnant of Israel) is in hiding is three and a half years, or FORTY-TWO months. It is the same FORTY-TWO months the holy city (Jerusalem) shall be trodden down. It is the same FORTY-TWO months the beast will continue in his beastly power. Christ, at His SECOND COMING to the earth, will destroy the man of sin, who is the beast. (II Thess. 2:4-8 and Revelation 19:11-21) This connects our Lord's coming with the number FORTY-TWO.

Our Lord's return to the earth at the end of the FORTY-TWO months of the reign of the beast must not be confused with His appearing in the air to catch away His saints.

In II Kings 2:23-25 we have the incident when two she bears came out of the woods, and tare FORTY and TWO children. Here the little children were the infidel young men of Bethel who were worshippers of the golden calf instead of Jehovah. It was God who sent the bears, and we have to believe that the offenders were worthy of such judgment. The term "bald head" had no special reference to the lack of hair, according to some authorities; it signified a worthless fellow. It was a term of contempt. Here it was equal to blasphemy of God for the young men mocked Elisha as a prophet of Jehovah, in contemptuous allusion to the translation of Elijah, which they no doubt denied and made fun of. The idea seems to be "Go up (be translated) like Elijah, you worthless fellow"!

Here again, FORTY-TWO is associated with the coming of Christ or the translation of the Saints that shall take place at the Rapture or the first phase of the Second Coming. This number is used FOURTEEN times in the Bible which reveals the DELIVERANCE of God's people at the Coming of Christ.

FORTY-FIVE
PRESERVATION

FORTY-FIVE seems to mean PRESERVATION. This number is found only a few times in the Bible but each time it reveals the power of God in preserving His saints.

In Genesis 18:28, when God is getting ready to destroy the wicked city of Sodom you will remember that Abraham interceded before God for the city. Note the Scriptures in Genesis 18:23-33,

> "Peradventure there shall lack five of the fifty righteous, wilt thou destroy all the city for lack of five? And he said, if I find there FORTY and FIVE, I will not destroy it."

You will note that God said he would preserve the city if there were FIFTY, FORTY-FIVE, FORTY, THIRTY, TWENTY or TEN righteous people in the city.

FIFTY is the number of the Holy Spirit. God is willing through His Spirit show mercy on the righteous; FORTY is the number of TESTING. The city is being tested. THIRTY is the number of the BLOOD and the HIGH PRIEST who offers the sacrifice upon the altar and who intercedes on the behalf of his people. TWENTY is the number of REDEMPTION that shall come if righteousness is found and TEN is the number of TESTIMONY, a testimony that shall be borne if TEN righteous are found.

All the numbers mentioned are even numbers save FORTY-FIVE. Here is PRESERVATION in the midst of sin.

In Joshua 14:10 the Bible says,

> "And now, behold, the Lord hath kept me alive, as he said, these FORTY and FIVE years, even since the Lord spake this word unto Moses, while the children of Israel

> wandered in the wilderness: and now, lo, I am this day fourscore and five years old."

God always keeps His Word with men. It would have been impossible for Caleb to die in all the past FORTY-FIVE years, by warfare or otherwise, since God had given him His word that he would live to inherit the place where he went as a spy. (Numbers 14:24)

God has preserved many men in times past because He gave His word, and He is yet to do this in the future, especially with the TWO witnesses who cannot die until they have finished their testimony. (Revelation 11:3-11) Thus we see that FORTY-FIVE is the number of PRESERVATION.

FIFTY
HOLY SPIRIT

FIFTY is the number connected with the Holy Spirit and His work. The Holy Spirit was poured out on the day of Pentecost, which was FIFTY days after the resurrection of Christ. In Leviticus 23:9-16 there is a beautiful and enlightening passage on the resurrection of Christ and the coming of the Spirit FIFTY days after His resurrection.

> "And the Lord spake unto Moses, saying, Speak unto the children of Israel, and say unto them, when ye be come into the land which I give unto you, and ye shall reap the harvest thereof, then ye shall bring a sheaf of the FIRSTFRUITS of your harvest unto the priest: and he shall wave the sheaf before the Lord, to be accepted for you: on the morrow after the Sabbath (that is, the FIRST day of the week) the priest shall wave it." (Leviticus 23:9-11)

This is a picture of the resurrection of Christ, which took place on the FIRST DAY of the week, or the morrow after the Sabbath.

> "In the end of the sabbath, as it began to dawn toward the FIRST day of the week, came Mary Magdalene and the other Mary to see the sepulchre. And, behold, there was a great earthquake: for the angel of Lord descended from heaven, and came and rolled back the stone from the door, and sat upon it." (Matthew 28:1-2)

Then the angel said to the women,

> "Fear not ye: for I know ye seek Jesus,

> which was crucified. He is not here: for He is risen." (Matthew 28:5-6)

Then in I Corinthians 15:20 it is said,

> "But now is Christ risen from the dead, and become the FIRSTFRUITS of them that slept."

Thus, the waving of the FIRSTFRUITS of the harvest on the day after the Sabbath was a picture of the Resurrection of Christ.

Now, go back to the passage in Leviticus, and see the outpouring of the Spirit typified.

> "And ye shall count unto you from the morrow after the sabbath, from the day that ye brought the sheaf of the wave offering; seven sabbaths shall be complete: even unto the morrow after the seventh sabbath shall ye number FIFTY days; and ye shall offer a new meat offering unto the Lord." (Leviticus 23:15-16)

This is a picture of the outpouring of the Spirit who came on Pentecost, FIFTY days after Christ arose from the dead.

Now go back to an examination of the Tabernacle in the Old Testament. There were FIFTY taches of gold connected with the linen curtains. There were TEN linen curtains, fastened together, FIVE in one group, and FIVE in another group. These TWO groups of FIVE each were coupled together with the FIFTY taches. (Exodus 26:1-6) The taches came over the veil. (Exodus 26:33) Which represents the crucifixion of Christ. (Hebrew 10:19-20) This connects the work of the Holy Spirit with the work of the Cross, and with the GRACE of God, represented with the number FIVE.

The court of the Tabernacle was FIFTY cubits wide. (Exodus 26:12-13) In the west end there were TEN pillars. Between these TEN pillars there were NINE spaces in the hanging that was FIFTY cubits long. NINE represents the FRUIT OF THE SPIRIT. The FRUIT of the SPIRIT (9) comes through the work of the HOLY SPIRIT (50), even as the NINE spaces are found in the curtain of FIFTY cubits length.

The number FIFTY is also connected with Israel's return and restoration to their land which God gave them. The FIFTIETH year was a year of Jubilee unto the people of Israel. It was a year when every man returned and repossessed any land that he may have had to sell because of debts.

> "And ye shall hallow the FIFTIETH year, and proclaim LIBERTY throughout all the land unto the inhabitants thereof: it shall be a JUBILEE unto you: and ye shall return every man to his possession, and ye shall return every man unto his family." (Leviticus 25:10)

The final and complete return of the Israelites unto their land is connected with their receiving the Spirit.

> "Therefore, thus saith the Lord God: Now will I bring again the captivity of Jacob, and have mercy upon the WHOLE house of Israel (twelve tribes), and will be jealous for my holy name; after they have borne their shame, and all their trespasses against me, when they dwelt safely in their land, and none made them afraid. When I have brought them again from the people, and gathered them out of their enemies' lands, and am sanctified in them in the sight of many nations; then shall they know that I am the Lord their God, which caused them to be led into captivity among the heathen: but I have gathered them unto their OWN LAND, and have left NONE of them ANY MORE there. Neither will I hide my face any more from them: for I have POURED OUT MY SPIRIT upon the house of Israel, saith the Lord God." (Ezekiel 39:25-29)

This teaches a return of the Israelites to their own land that will not leave any of them in the Gentile countries. God speaks about a time when He will not leave any of them in any foreign land. This will take the last living Jew out of the GENTILE countries and put them in their own land again. And they shall not ANY MORE be scattered among the Gentile countries. As long as one Jew still walks the streets of Gentile countries that prophecy is yet to be fulfilled. Then the Lord will pour out His Spirit upon them when they shall have all been regathered. The 29th verse shows exactly that.

Thus, the FIFTIETH year, a year of JUBILEE, when every man returned to his possession, was a picture of the time when all Israel will have returned to their possessions: the land God gave to that people. And the SPIRIT (50) will be poured out upon them. For Israel that will be a time of JUBILEE of GREAT JOY.

> "And the ransomed of the Lord shall return, and come to Zion (their possession) with songs and everlasting joy (Jubilee) upon their heads: they shall obtain joy and gladness, and sorrow and sighing shall flee away." (Isaiah 35:10)

This wonderful teaching of the HOLY SPIRIT is revealed again in Numbers 4. Read the 3rd, 23rd, 30th, 35th, 39th, and 43rd verse.

The work age of the priest to serve in the Tabernacle was between the ages of THIRTY and FIFTY. Some TWENTY years a priest was to administer at the altar the blood sacrifice for the REDEMPTION of sins. TWENTY is the number for REDEMPTION. THIRTY is the number of the BLOOD OF CHRIST which was typified in the altar sacrifices and which was to be offered by the Priest. FIFTY is the HOLY SPIRIT which comes as the result of the BLOOD OF CHRIST being offered upon the altar.

Jesus said in John 16:7,

> "Nevertheless I tell you the truth; It is expedient for you that I go away: for if

> I go not away, the Comforter will not come unto you; but if I depart, I will send him unto you."

Also in Hebrews 9:11,

> "But Christ being come an high priest of good things to come; by a greater and more perfect tabernacle, not made with hands, that is to say, not of this building; Neither by the blood of goats and calves, but by his own blood he entered in once into the holy place, having obtained eternal redemption for us."

Jesus said the HOLY SPIRIT could not come until He returned to the Father and there to act as our mediator or HIGH PRIEST and as He placed HIS BLOOD (30) on the altar for the atonement (20) for our sins, then the HOLY SPIRIT (50) would come to bring COMFORT to the hearts of the believers in Christ. And on the FIFTIETH day the Spirit came to take up His office work during these days of GRACE. How marvelous is the Word of God! This number is found 154 times in the Bible.

SIXTY
PRIDE

SIXTY seems to stand for PRIDE. The image which Nebuchadnezzar set up was SIXTY cubits (90 feet) high. (Daniel 3:1) PRIDE prompted him to erect this image. He dreamed of a great image (chapter 2) whose head was of gold. Daniel told him that he (Nebuchadnezzar) was that head of gold. (Daniel 2:36-38) This, along with his greatness, filled him with PRIDE, as recorded in chapter 4, verse 30.

> "Is not this great Babylon, MY power, and for the honor of MY MAJESTY."

While he was yet speaking a voice came from heaven saying,

> "O king Nebuchadnezzar, to thee it is spoken; the kingdom is departed from thee." (vs. 3)

For seven years he was deprived of his reason and he was made to eat grass with the beasts of the field. This was God's punishment upon him for his PRIDE. This record follows immediately after his erection of the great image, SIXTY cubits high.

SIXTY is SIX times TEN, the numbers for Satan and the LAW. Satan fills those with PRIDE and boasting who are under the LAW.

> "Where is boasting then? It is excluded. By what LAW? Of works? Nay: but by the law of faith." (Romans 3:27)

This number is mentioned 14 times in the Word of God.

SIXTY-SIX
IDOL WORSHIP

SIXTY-SIX is the number connected with IDOL WORSHIP. The image which Nebuchadnezzar erected to be worshipped was SIXTY cubits high, and SIX cubits broad.

> "Nebuchadnezzar the king made an image of gold, whose height was threescore (60) cubits, and the breadth thereof six cubits: he set it up in the plain of Dura, in the province of Babylon." (Daniel 3:1)

The verses that follow this show that Nebuchadnezzar commanded all people to worship his golden image, or be cast into the burning firey furnace. (Daniel 3:23) This connects SIXTY-SIX with IDOL WORSHIP.

Jeremiah prophesied that Judah would be carried into Babylonian captivity because of their IDOLATRY He said unto the people,

> "The Lord hath sent unto you all His servants the prophets, but ye have not harkened, nor inclined your ear to hear. They said, Turn again now, every one from his evil way . . . and go not after OTHER GODS to serve them and to worship them, and to provoke me to anger with the works of your hands; and I will do you no hurt. Yet ye have not harkened unto me, saith the Lord . . . Therefore thus saith the Lord of hosts, because ye have not heard my words, behold, I will send and take all the families of the north, saith the Lord, and Nebuchadnezzar the king of Babylon, my servant, and will bring them against this land and against the inhabitants thereof . . . and this whole land shall be a desolation, and an astonishment; and these nations shall serve the king of Babylon, SEVENTY years. (Jeremiah 25:4-11)

FROM this it is seen that Judah went into Babylonian captivity SEVENTY years because of IDOL WORSHIP. It has already been seen that the number SIXTY-SIX is connected with IDOL WORSHIP. Now see how their IDOL WORSHIP and the SEVENTY years of Babylonian captivity was forecast in the numbers in Jacob's family that went into Egypt.

> "All the souls that came with Jacob into Egypt, which came out of his loins, besides Jacob's sons' wives, all the souls were THREESCORE and SIX (66); and the sons of Joseph, which was born to him in Egypt, were two souls; all the souls of the house of Jacob, which came into Egypt, were THREE-SCORE and TEN." (Genesis 46:26-27)

SIXTY-SIX of Jacob's descendants went with him into Egypt. Jacob himself, Joseph, and Joseph's two sons made SEVENTY in all who were in Egypt. Their sojourn ended in bondage. In the number SIXTY-SIX, the number found in verse 26, there is foreshadowed the reason why Judah would go into Babylonian captivity, that is for IDOL WORSHIP, represented by SIXTY-SIX. The SEVENTY in verse 27 forecasts the duration of the Babylonian captivity, SEVENTY years. Was this an accident? If so, why did all those numbers exactly correspond to the number for IDOL WORSHIP, and to the number of years they were in BABYLONIAN captivity? Jeremiah's prophecy was spoken about eleven hundred years after Jacob went into Egypt.

It was in the reign of Josiah, in the THIRTEENTH year, that Jeremiah was called to prophesy against Judah's evil rebellious ways. (Jeremiah 1:2 and 25:3) This is the number for REBELLION. This number is mentioned ONLY SIX times in the Bible, the number SIXTY-SIX meaning IDOL WORSHIP.

SEVENTY

UNIVERSALITY - - ISRAEL AND HER RESTORATION

SEVENTY is the number of UNIVERSALITY. The descendants of Shem, Ham, Japheth, who repopulated the earth after the flood were SEVENTY in number. (Genesis 10)

SEVENTY persons of Israel's seed came down into Egypt at Joseph's request. (Genesis 46:27; Exodus 1:5) Joseph was of the house of Jacob, and came into Egypt from Canaan. Adding him, his two sons and Jacob, his father to the SIXTY-SIX out of Jacob's loins (Genesis 46-26-27) accounts for the THREESCORE and TEN or SEVENTY of the house of Jacob. God used this group to form the nation of Israel.

SEVENTY years Israel lived in exile. (Jeremiah 25:11; 29:10) This was the whole length of the captivity of the Jews in Babylon. After this Babylon was to be overthrown and the Jews were to return to their own land. (Jeremiah 25:11-12; Daniel 9:2; Zechariah 7:5) Nebuchadnezzar began his seige of Tyre in the first year of his reign and from then to the taking of Babylon by Darius and Cyrus was SEVENTY years. This was the length of the Babylonian monarchy.

God gave Moses the command to choose SEVENTY elders to assist him in his duties. (Numbers 11:16) Here they were commissioned and anointed with the Holy Spirit to take part in the responsibility of Moses as head of the nation. (vs 16, 24-25). Some say, this was the beginning of the Sanhedrin. These were the elders of Israel. (Vs. 16).

Jesus chose SEVENTY disciples to go out in His name. (Luke 10)

SEVENTY is the number connected with God's punishment of ISRAEL for her disobedience. II Chronicles 36:20-21 reveals Judah's Babylonian captivity was due in part to their refusal to keep the sabbath year.

> "And them that escaped from the sword carried he away to Babylon; where they were

> servants to him and his sons until the reign of the kingdom of Persia: to fulfill the word of the Lord by the mouth of Jeremiah, until the land had enjoyed her sabbaths: for as long as she lay desolate she kept the sabbath, to fulfill THREESCORE and TEN."

In Leviticus 25:1-4 God commanded Israel to keep a sabbath year every SEVENTH year. Because they had failed to keep SEVENTY sabbaths the land had to lie idle SEVENTY years. Their idolatry had caused them to also disobey God's law concerning the sabbath year SEVENTY times.

God permitted Judah to go down completely and suffer the bitterness of shame and despair because there was no remedy for them otherwise. The condition continued for SEVENTY years, and the land enjoyed the Sabbaths. (Jeremiah 25:9) What the Israelites would not do willingly, God did by exile.

Since a sabbatical year was supposed to be observed every SEVENTH year then SEVENTY sabbatical years would reach over a period of 490 years, or SEVENTY times SEVEN. This throws light on our Lord's statement to Peter in Matthew 18:21-22 and on Daniel's prophecy in Daniel 9:24-27.

Peter asked the Lord saying,

> "How oft shall my brother sin against me, and I forgive him? Till seven times."

Then Jesus said,

> "I say not unto thee, until seven times, but until SEVENTY times SEVEN."

Then the Lord said to Daniel,

> "SEVENTY weeks (or SEVENS) are determined upon thy people and upon the holy city, to finish the transgression, and to make an end of sins, and to make reconciliation for iniquity, and to bring in everlasting

> righteousness, and to seal up the vision and the prophecy, and to anoint the most Holy." (Daniel 9:24)

Notice these SEVENTY weeks, or SEVENTY times SEVEN years, were to bring an end to the sins and trangressions of Daniel's people, the Israelites. The application of the Bible system of numbers will help to understand this passage which has been the source of much controversy.

First of all, notice that these SEVENTY weeks, or SEVENTY times SEVEN years equal the duration of times in which Israel failed to observe her sabbatical years. SEVENTY times SEVEN equal 490 years, the duration of time in which a sabbatical year was not observed. Since this period of time had to do with a land which God gave to Israel one would expect to find the SEVENTY weeks, or 490 years connected with God's blessings upon the land of Israel. Israel sinned in failing to observe God's law concerning God's sabbatical year. Consequently when the full time of Israel's punishment for her sins had run its course one would expect to find God's blessings upon Israel's land again. In this connection it is well to study the 85th Psalm.

> "Lord, thou hast been favorable unto THY LAND: thou hast brought back the captivity of Jacob. Thou hast forgiven the iniquity of thy people, thou hast covered ALL their sin. Selah. Thou hast taken away all thy wrath: thou hast turned thyself from the fierceness of thine anger." (Psalm 85: 1-3)

Verse 12 reads, "Yea, the Lord shall give that which is good; and our land shall yield her increase."

These verses teach that Israel's return from captivity, and the forgiveness of Israel's sins is connected with God's blessings upon the land, the same land where Israel refused for 490 years, or SEVENTY times SEVEN years, to observe a sabbath year. That would be SEVENTY sabbath years that were not kept.

> "The Lord said to Daniel, SEVENTY weeks (70 x 7) are determined upon (1) thy people and upon (2) thy holy city (Jerusalem), (3) to finish the trangression, and (4) to make an end of sins (the sins of Israel, Daniel's people), and (5) to make reconciliation for iniquity, and (6) to bring in everlasting righteousness, and (7) to seal up the vision and prophecy, and (8) to annoint the most Holy." (Daniel 9:24)

In the above there are EIGHT things connected with the SEVENTY weeks that are determined upon Israel, Daniel's people. EIGHT is the number for the new birth, and points to a time when Israel's transgression has been finished; her sins have been ended, when the people have been reconciled with God; when the prophecies concerning them have been fulfilled; when they have been born again; when they have been anointed with the Spirit from on high; and when they have been brought back from captivity; and when God's blessings is once more poured out upon her land.

One thing that must be considered in connection with Daniel's prophecy about the SEVENTY weeks (70 x 7) years is the prayer of Daniel that just preceded this prophecy.

> "O Lord hear: O Lord forgive; O Lord harken and do: defer not for thine own sake, O my God: for the city and thy people are called by thy name. And whiles I was speaking and praying and confessing my sin and sin of my people Israel . . . yea, whiles I was speaking in prayer, even the man Gabriel . . . touched me about the time of the evening oblation. And he informed me . . . and said, O Daniel, I am now come forth to give thee skill and understanding. At the beginning of thy supplications the commandment came forth, and I am come to shew thee, for that art greatly beloved: therefore understand the matter, and consider the

> vision. SEVENTY WEEKS are determined upon thy people." (Daniel 9:18-24)

Thus the vision of the SEVENTY WEEKS was given in answer to Daniel's prayer for God to forgive Israel's sins. For this reason one should expect Israel's sins to be forgiven and Israel to be restored at the end of the SEVENTY WEEKS.

Another thing to be considered is that there are THREE divisions to the SEVENTY weeks of Daniel: SEVEN weeks, THREESCORE and TWO (62) weeks, and ONE week. (Daniel 9:24-27) There are SEVEN weeks or SEVENS (49) years until the going forth of the commandment to build the city. This probably refers to the time of Nehemiah, who rebuilt the walls of Jerusalem in troublous times. Daniel said,

> "The street shall be built again, and the WALL, even in troublous times." (Nehemiah 4:6)

The rest of the chapter tells of the opposition the Jews had while building the wall, and how they continually had their swords and spears with them as they worked. (Verses 7-23)

SIXTY-TWO more weeks of the SEVENTY ends with Israel's rejection of Christ, and His crucifixion.

> "After THREESCORE and TWO weeks shall Messiah be cut off, and not for himself." (Daniel 9:26)

This leaves ONE more week, or a period of SEVEN YEARS, to finish the SEVENTY. This is the THIRD of the THREE divisions of Daniel's SEVENTY WEEKS, viz: SEVEN weeks, SIXTY-TWO weeks, and ONE week. THREE is the number for the RESURRECTION. Therefore, according to the Bible rule of numbers, one would expect the THIRD division of Daniel's SEVENTY weeks, or the ONE week, (SEVEN YEARS), to come beyond the RESURRECTION OF THE SAINTS. This is confirmed by the closing words of the Lord to Daniel,

> "But go thy way till the end be: for thou shall rest (die), and stand in thy lot (be raised again) at the end of the days." (Daniel 12:13)

To understand why the SIXTY-NINE (7 plus 62) weeks ended when Christ was crucified and the last week is yet future, it must be kept in mind that Israel was cut off when Christ was rejected and crucified, to be grafted in later on. Just a few days before His crucifixion Jesus wept over Jerusalem and said,

> "Behold your house is left unto you desolate." (Matthew 23:37-30)

Israel's time clock stopped at the end of SIXTY-NINE weeks. Then Jesus said,

> "For I say unto you, ye shall not see me henceforth, till ye say, Blessed is he that cometh in the name of the Lord." (Verse 39)

Israel is now cut off from her tame olive tree. (Romans 11:11-25) Temporary blindness has come upon them which is to last until the fullness of the Gentiles has come in. (Verse 25) When the fullness of the Gentiles has come in, then the time will have come for Israel to be grafted in again. Israel's time clock will start once more and the SEVENTIETH WEEK, or last SEVEN years, will come in.

During the last 2000 years, during this church age, the Lord is not dealing with the ISRAELITES AS A NATION. He is now calling out His Bride. When the Body of Christ is completed, the rapture will take place, and then the Lord will begin the last SEVEN years of Daniel's prophecy for the ISRAELITES. This period is called the TRIBULATION in the Book of Revelation.

Daniel 9:27 shows the LAST WEEK, the SEVENTIETH, will be divided into TWO periods of THREE and an HALF YEARS each.

> "And he shall confirm the covenant with many for ONE week: and in the midst of

> the week he (the little horn or Beast) shall cause the sacrifice and the oblation to cease, and for the overspreading of abominations he shall make it desolate, even until the consummation (the end of the SEVENTIETH WEEK) and that determined shall be poured upon the desolate."

The last half of this week will be the FORTY-TWO months the holy city (Jerusalem) shall be trodden down, according to the prophecy of Revelation 11:1-2. It will be the FORTY-TWO months of the universal power and persecution of the beast of Revelation 13:4-8. (You will want to read my book, RICHES OF REVELATION, in which I go into detail in these chapters of Revelation.)

At the close of the FORTY-TWO months of the cruel reign of the Anti-Christ, or the Beast, Christ will return in glory to the earth and the Beast will meet his doom. (Revelation 19:11-21) This will end Israel's SEVENTIETH WEEK. It will usher in the THOUSAND YEARS reign, and Israel's wanderings will be over. (Revelation 20:1-6)

Isaiah said to Israel,

> "Thy sun shall no more go down; neither shall thy moon withdraw itself: for the Lord shall be thine ever lasting light, and the days of thy mourning shall be ended. Thy people shall be ALL righteous: they shall inherit the land forever, the branch of my planting, the work of my hands, that I may be glorified . . . I the Lord will hasten it in his time." (Isaiah 60:20-22)

When this is accomplished Israel's transgression will be finished, and an end will have been made to their sins: reconciliation will have come for them; everlasting righteousness will have been brought in; and the vision and prophecy concerning them will have reached its fulfillment, according to Daniel's prophecy of the SEVENTY WEEKS determined upon that people. Unaerstand this and the teachings of the Book of Revelation becomes easier to understand.

The opponents of the Pre-millennial position on Daniel's SEVENTY weeks stop the SEVENTY weeks with the destruction of Jerusalem by General Titus in 70 A. D. They end them with Israel still in sin and unbelief. They ignore Daniel's prayer for forgiveness for Israel. They ignore the end of Israel's transgression and sins about which Daniel speaks. They ignore the reconciliation Daniel foretells for his people. They ignore Israel's restitution. They ignore the part concerning the land of Israel, which was connected with the sabbath years, plays in all this prophecy. They do not take into consideration a forgiveness of sins for Israel after her punishment has ended. They pay no attention to the Bible rule of numbers. Jesus connected SEVENTY times SEVEN with forgiveness of sins in Matthew 18:21-22. Israel failed to keep SEVENTY sabbath years. (I Chronicles 36:21) A sabbath year came every SEVENTH year. (Note the SEVENTH SABBATH in Leviticus 25)

> "In the SEVENTH year shall be a sabbath of rest unto the land." (Leviticus 25:4)

Thus Israel sinned against the law of the sabbath for 490 years, or SEVENTY times SEVEN years. When this period of Israel's time has run her sins will be pardoned, and they (Israel) will be restored to the land, and the land will be blessed.

SEVENTY is mentioned 61 times in the Bible.

ONE HUNDRED
GOD'S ELECTION OF GRACE - - CHILDREN OF PROMISE

ONE HUNDRED is the number that stands for GOD'S ELECTION OF GRACE, or THE CHILDREN OF PROMISE. Isaac, a child of promise, was born when his father was a HUNDRED years old.

> "And Abram was an HUNDRED years old when his son Isaac was born unto him. (Genesis 21:5)

Isaac was a type of the children of promise, or promised seed.

> "Now we, brethren, as Isaac was, are the children of promise." (Galatians 4:28)

> "In Isaac shall thy seed be called. That is, they which are children of the flesh, these are not the children of God: but the children of the promise are counted for the seed." (Romans 9:7-8)

The number ONE HUNDRED is connected with Isaac's sowing and reaping, and GOD's blessing.

> "Then Isaac sowed in that land, and received in the same year an HUNDRED fold: and the Lord BLESSED him. (Genesis 26:12)

Isaac received this HUNDRED FOLD in the harvest time. In His parable about the tares Jesus said,

> "The harvest is the end of the world." (Matthew 13:39)

Then the Lord will gather His wheat (harvest) into His

barn. (Matthew 13:30) This reaping will be at the end of the Kingdom age.

Jesus likened those whom He would save unto an HUNDRED sheep, gathered into a fold.

> "For the Son of man is come to save that which was lost. How think ye? If a man have an HUNDRED sheep, and one of them be gone astray, doth he not leave the ninety and nine, and goeth into the mountains, and seeketh that which is gone astray?" (Matthew 18:11-12)

He said,

> "Other sheep I have, which are not of this fold: them also I must bring, and they shall hear my voice; and there shall be ONE FOLD and ONE HUNDRED. (John 10:16)

By this Jesus compares His OWN to SHEEP, which shall be gathered into ONE FOLD. In the parable in Matthew 18:11-12 He used an HUNDRED sheep to illustrate those whom He would save.

Now, having seen that our Lord compared His own to sheep gathered into ONE FOLD, let us apply this to the Tabernacle. The court of the Tabernacle was an enclosure round about the Tabernacle and its furniture, like unto a SHEEP FOLD. That court was an HUNDRED cubits long, FIFTY cubits broad, EVERY WHERE, and the height was FIVE cubits of fine twined linen, and their sockets of brass. (See Exodus 26:18)

In the HUNDRED cubits in length the doctrine of ELECTION is taught. In the FIVE cubits is GRACE. In the length and height together is GOD'S ELECTION OF GRACE. The FIFTY cubits in the width of the court teach of the Holy Spirit being given to ALL of God's ELECTION OF GRACE. In speaking of the Holy Spirit Peter said,

> "The promise is unto you, and to your children, and to all that are afar off, even AS

MANY AS THE LORD GOD SHALL CALL." (Acts 2:39)

There were ONE HUNDRED sockets of silver in the Tabernacle, which were made of ONE HUNDRED talents of silver, which money had been given in REDEMPTION.

"And the silver of them that were numbered. (Exodus 30:12-14) of the congregation was an HUNDRED talents, and a thousand seven hundred and fifteen shekels, after the shekel of the sanctuary." (Exodus 38:25)

"And of the HUNDRED talents of silver (redemption money of Exodus 30:12-14) were cast the sockets of the sanctuary, and the sockets of the veil; and HUNDRED sockets of the HUNDRED talents, a talent for a socket." (Exodus 38:27)

Thus the total number of talents of redemption money (silver) that was used in the Tabernacle was ONE HUNDRED, and the silver sockets were ONE HUNDRED. This was the sum total enclosed by the court and in the Tabernacle itself. Therefore this ONE HUNDRED is expressive of the full number of the SAVED, or ALL those who will be saved.

Let not the reader get the idea that only ONE HUNDRED will be saved. The number ONE HUNDRED is that number that signifies the ELECT or the SAVED. Only God Himself knows how many that will be.

But there was some REDEMPTION money over and above that which went into the HUNDRED sockets of silver in the Tabernacle. There were ONE THOUSAND, SEVEN HUNDRED and SEVENTY-FIVE shekels above the HUNDRED talents. This went into the making of the hooks and fillets for the pillars and chapters of the court. (Exodus 38:28)

The hooks of the pillars in the veil and the door of the Tabernacle were made of gold. (Exodus 26:31 and 37) This extra amount of silver on the pillar around the court teaches an offer of redemption to all people.

But in the HUNDRED silver sockets in the Tabernacle, and in the court being ONE HUNDRED cubits long, there is a lesson picturing only those who will avail themselves of God's offered Redemption. They are the HUNDRED sheep in the fold.

Perhaps the reader has noticed that the number ONE HUNDRED occurs twice in the court, once in the hanging for the south side, and once for the hanging for the north side. Those hangings of ONE HUNDRED cubits, hung upon the TWENTY pillars on each side. TWENTY represents REDEMPTION. (See Exodus 27:9-11) There is a two fold REDEMPTION for the SAVED: the REDEMPTION FOR THE SOUL, and the REDEMPTION FOR THE BODY. The sides are equal, teaching that as many as receive REDEMPTION for their souls will likewise receive the REDEMPTION for their bodies.

> "Whom he justified, them he also glorified." (Romans 8:29)

There were also ONE HUNDRED taches in the curtains of the tabernacle.

> "In the linen curtains there were FIFTY taches of gold." (Exodus 26:6)

In the goat's hair curtains there were FIFTY taches of brass. (Vs 11) This makes ONE HUNDRED TACHES, or two times FIFTY. FIFTY is the number for the Holy Spirit. This teaches that the Spirit is given to every child of promise.

> "The promise is to you, and your children, and to all that are afar off, even as many as the Lord our God shall call." (Acts 2:39)

The Holy Spirit not only witnesses to salvation (Romans 8:16), but He guarantees the redemption of the body. (Read Ephesians 1:13; 4:30; II Corinthians 5:5)

ONE HUNDRED NINETEEN

THE RESURRECTION DAY - - LORD'S DAY

In this number we have a striking instance confirming the fact that even in the numbering of chapters and verses, which was done in comparatively modern times, God over ruled the minds of those who did it, so that the same hidden meanings of numbers be carried out. I suppose that everyone has noticed how frequently the same number of chapter and verse brings out an important truth, and these verses seem to be specially marked in this way to draw our attention to them.

Who does not know John 3:16, that wondrous Message of Salvation, then Matthew 3:16 which gives us the descent of the Holy Spirit like a dove upon the Son of God, to anoint Him for that glorious work of Salvation. Luke 3:16 gives witness to the fact that this Anointed Messenger should be the One Who would baptize "with the Holy Ghost and with fire" Acts 3:16 emphasizes that "faith in His name" works wondrous miracles of grace. I Corinthians 3:16 tells believers that they are "the Temple of God" through the indwelling of the Spirit of God that reveals Salvation. Ephesians 3:16 tells us that it is the Holy Spirit that strengthens "the inner man" who has Salvation. Colossians 3:16 gives the occupation and behaviour of the believer who professes God's Salvation. II Thessalonians 3:16 tells that "the Lord of peace Himself" gives peace to His disciples who are in Salvation. I Timothy 3:16 gives the whole Gospel in a nutshell, and II Timothy 3:16 lays down that "all Scripture is given by inspiration of God." While I John 3:16 speaks of "the love of God" in laying down "His life for us." THREE speaks of Divine perfection, while SIXTEEN, being the double of EIGHT, (new birth) has its significance of LOVE, only intensified.

This same truth is found in the number SEVENTEEN. There are SEVENTEEN verses in the first chapter of Leviticus, which gives the first detailed instructions for the Burnt Offering, "an offering made by fire, of a sweet savour unto the Lord," as we are told three times over, emphasizing its DIVINE VICTORY. This is a type of the Lord Jesus offering Himself an Acceptable Offering to God the Father, by which He became the Accepting-Place of sinners. This is our VICTORY.

The SEVENTEENTH chapter of Genesis is devoted to the Everlasting Covenant which God made with Abraham through the rite of circumcision, by which the Jews were set apart as God's earthly people and through them VICTORY would come from their seed.

The SEVENTEENTH chapter of Exodus gives us the story of the smitten rock out of which came water for the thirsty multitude, "they drank of that spiritual rock that followed them: and that rock was Christ."

In the SEVENTEENTH chapter of Leviticus we read that "the life of the flesh is in the blood: and I have given it to you upon the altar, to make an atonement for your souls: for it is the blood that maketh atonement for the soul."

In the SEVENTEENTH chapter of Numbers, the rod of Aaron buds and "bloomed blossoms, and yielded almonds," telling us of life only in Christ. This is our VICTORY.

> "He that abideth in Me, and I in him, the same bringeth forth much fruit." (John 15:5)

The SEVENTEENTH Psalm is one that speaks of God in the words "Thine eyes", "Thy lips", "Thine ear", "Thy right hand", "Thy wings", "Thy face", and "Thy likeness", with which we shall be satisfied when we awake!

The SEVENTEENTH chapter of John, which is devoted to the great High-Priestly prayer of our Lord for His believers. Here our VICTORY is assured in the prayers of our High Priest.

The longest chapter in the Bible is the 119th Psalm, which contains no less than 176 verses. It is more than double the length of any other chapter, and it is devoted wholly to one subject, the Word of God! Moreover it is in the very heart of the Bible, the actual middle verse of the Word, being the EIGHTH verse of the preceding Psalm. The Hebrews call it the "Great Alphabet", because every verse begins with a letter of the Hebrew alphabet, one section of verses all beginning with aleph, another section with beth, and so on. Thus does the Holy Spirit teach us that the alphabet of spiritual knowledge is contained in God's Word.

If we neglect to learn God's great alphabet, we shall not come to a knowledge of spiritual life, or, having been born again, we shall make no progress, shall not grow in grace, unless we devote time to the study of that same alphebet. It is the alphabet of all true spiritual understanding.

Now we have seen that TEN is the number of TESTIMONY, LAW AND RESPONSIBILITY, and the next lesson taught us in this Psalm, is through the fact that TEN special words are used, one of which occurs in every single verse of this long Psalm, except the 122nd verse, which contains instead, the very Name of Christ as our "Surety", "a Surety of a better Testament", as we learn in Hebrew 7:22.

> "By so much was Jesus made a surety of a better testament."

These TEN words are:: WAY, TESTIMONIES, PRECEPTS, COMMANDMENTS, SAYING, LAW, JUDGMENT, RIGHTEOUSNESS, STATUTES, AND WORD. It will be noted that they are all descriptive of God's Word, and their number TEN, tells us that only through God's Word can we obtain a TESTIMONY and RESPONSIBILITY to the world.

It is the same lesson that the TEN COMMANDMENTS teach us. These TEN words occur 204 times in all or SEVENTEEN times TWELVE, that is the number of VICTORY multiplied by the number of GOVERNMENTAL PERFECTION. It is by the perfect government of our Master submitted to in our lives that VICTORY is wrought out in us.

Now we come to the number of the Psalm itself.

Why is it the 119th Psalm and not some other number? Many times spiritual truths are hidden in a compound number and the answer can be found in its simplest factors. But ONE HUNDRED AND NINETEEN has only TWO factors, that is there are only TWO numbers by which it is divisible, and these are SEVENTEEN and SEVEN! SEVEN times SEVENTEEN is ONE HUNDRED AND NINETEEN. What does this teach us? Probably the primary lesson is that both VICTORY and SPIRITUAL PERFECTION are found in God's Word.

But a deeper truth even than this seems to be hidden in this number, for we remember that there is one special day marked out by THREE epoch-making events given in the Bible, and that is the SEVENTEENTH DAY of the SEVENTH month! The wonderful day on which the Ark rested safely on the mountains of Ararat, after the storms of God's judgments had cleansed the earth.

Also this was the same day of that marvelous deliverance which God wrought for the children of Israel in bringing them through the Red Sea upon the dry ground, and over whelming their foes behind them in the returning waters. Both these exhibitions of God's mighty power were also designed as type-pictures or foreshadowings of that greatest Event of all, when the Crucified Saviour rose from the dead, the bars of death were broken, and witness was given by God that, He Who shed His precious blood for us, and poured out His soul unto death, was accepted as "the Propitiation for our sins," and had been raised from the dead, "for our justification". This happened on that SEVENTEENTH day of the SEVENTH month.

Thus we see that this ONE HUNDRED AND NINETEEN Psalm bears hidden in its very number the same glorious truth of salvation through faith in a Risen Saviour. And this Psalm which is all about the Word of God bears witness in its number to the one great subject of God's Word, the Living Word, a Saviour to come, Who should be "delivered for our offences", and "raised again for our justification".

But there is more even than this, for it will be noticed that while the number of the Psalm marks the day, the number of verses in each alphabetic section mark what happened on that great day, the day of RESURRECTION. For each section has EIGHT verses only, and EIGHT is the number of NEW BEGINNING or the NEW BIRTH. On that first SEVENTEENTH day of the SEVENTH month there were EIGHT persons saved in the Ark, and it was on the FIRST day of the week, the EIGHTH day as it were, that our Lord rose from the dead. And it was again on that EIGHTH day that the Holy Spirit came from heaven, through Whose gracious work in our hearts, we can walk in the RESURRECTION life day by day.

Thus this whole great Psalm is designed by the Holy Spirit to be a kind of kindergarten alphabet lesson to track the bedrock foundation truths, which God's Word in all its parts reveal to us, by such simple illustrations as the alphabet letters and numbers. How it strengthens one's faith in that wonderful revelation to see that in its very make-up the same glorious message of love from God to man is recorded.

Another instance of the use of these numbers in conjunction by the Holy Spirit is in Genesis 6, where the posterity of Shem, through whom the chosen race was to come, is given. Nahor is the SEVENTH descendant from Shem to be born after the Flood, and as soon as he bears the EIGHTH, Terah, Abraham's father, we are immediately told,

> "and Nahor lived after he begat Terah an hundred and nineteen years."

Thus, again are these two significant numbers brought together just prior to the birth of Abraham, in whom all families of the earth were to be blessed. And the Scripture, foreseeing that God would justify the heathen through faith, preached before the Gospel unto Abraham, saying, "In thee shall all nations be blessed." So Christ was in due time raised again for the justification of the lost who should believe on Him, on the SEVENTEENTH day of the SEVENTH month, on the EIGHT day of the week.

In the Greek word, "DEUTE", which means "Come", we have another example of the use of this number.

This word occurs exactly THIRTEEN times, and its gematria is 714, or SIX times ONE HUNDRED and NINETEEN. SIX speaks of man, and ONE HUNDRED NINETEEN, or SEVEN times SEVENTEEN, of how he can be separated from his sins (13) by faith in a Saviour, Who was raised from the dead on the SEVENTEENTH day of the SEVENTH month. The places where this word "come" is used makes a very interesting Bible study. (They are: Matthew 4:19; 11:28; 21:38; 22:4; 25:24; 28:6; Mark 1:17; 6:31; 12:7; Luke 20:14; John 6:29; 21:12; Revelation 19:12.)

In every case the passage refers to the Lord Jesus Christ, but in THREE instances the word is used by the enemies of Christ, who desire to make away with Him, and the last "Come" is the invitation to the judgment scene, when God's righteous wrath will fall on those who will not accept the "come" of mercy.

The first occurrence in Matthew 4:19 is translated,

> "Follow Me, and I will make you fishers of men."

It is literally, "Come behind", and shows us that if we would be used in His service we must be hidden behind Him. Our work is to exalt Christ, and for that the messenger must be hidden in Him.

The next place is the loving Saviour's invitation to the burdened sinner:

> "Come unto Me, all ye that labour and are heavy laden, and I will give you rest." (Matthew 11:28)

Because of His work on the cross, and of that glorious Resurrection on that glad SEVENTEENTH day of the SEVENTH month, the Saviour can give relief from the burden of sin, and sure hope of a joyful resurrection.

You will note that the Greek word "AGAPE" -- "love" -- occurs SEVEN times in John's Gospel and in the Epistle of John it is used SEVENTEEN times, thus linking up these TWO numbers, in John's writings, to bring out the truth that God's love to man is clearly revealed by what happened on that SEVENTEENTH day of the SEVENTH month, when the Lord Jesus Christ was raised for our justification!

The figure EIGHT too, significant of the NEW BIRTH or NEW BEGINNING, again comes in as usual, in the fact that there are only EIGHT miracles or "signs" recorded in John's Gospel, and the EIGHTH is the only miracle performed by our Lord after He "rose from the dead"! How clearly design is shown all through this marvelous Revelation of God to man.

Through man other instances of this number might be given, we will give but one more, and that a very interesting one. It is "KENTRON" -- "sting, prick". It only occurs FIVE times -- (Acts 9:5; 26:14; I Corinthians 15:55, 56; and Revelation 9:10). Its number is 595, or ONE HUNDRED AND NINETEEN times FIVE. Number FIVE is the number of God's GRACE. Note how the GRACE OF GOD is revealed through His Word.

Its significance is clearly seen in Acts, where it is translated "prick", and I Corinthians 15:55, where it is given as "sting". In Acts 9:5, it is easy to see what the "prick" was against which Paul was kicking. He was kicking against the fact which his conscience told him was true, that the Lord Jesus had risen from the dead on the SEVENTEENTH day of the SEVENTH month. He did not want to believe it, but he got no peace till he surrendered his sinful soul at the loving Saviour's Feet.

In I Corinthians 15:55, we read,

> "O death, where is thy sting?"

and in verse 56,

> "The sting of death is sin."

Because of the glorious Resurrection of the Lord Jesus so beautifully unfolded in this great chapter, death had been robbed of its sting! Those who believe in their hearts, "that God hath raised Him from the dead," will, by virtue of His atoning blood, never feel the "sting" of death, though they may feel its stroke. The "sting" has been drawn through His wonderful grace!

ONE HUNDRED TWENTY
DIVINE PERIOD OF PROBATION

ONE HUNDRED and TWENTY is made up of THREE FORTIES (3 x 40 equals 120). Applied to time it therefore signifies a DIVINELY appointed period of PROBATION. This truth is revealed in the days preceding the Flood:

> "And the Lord said, My spirit shall not always strive with man, for that he also is flesh: yet his days shall be an HUNDRED and TWENTY years." (Genesis 6:3)

When this number is applied to persons it points to a DIVINELY appointed number during a period of waiting.

> "And in those days Peter stood up in the midst of the disciples, and said, (the number of names together were about an HUNDRED and TWENTY). (Acts 1:15)

It is a factor in the number of those who returned from Babylon.

> "The whole congregation together was FORTY and TWO thousands THREE HUNDRED and THREESCORE.

This number 42,360 is 120 x 353.

It is also a factor on the number of the men who went up out of Egypt, 600,000, being 120 x 5000.

It is a factor of the 144,000 who will be sealed from the Twelve Tribes of Israel to go unscathed through the great Tribulation, 144,000 being 120 x 1200. (Revelation 7)

This number is found only a few times in the Bible.

ONE HUNDRED AND FORTY-FOUR

THE SPIRIT GUIDED LIFE

This number is the square of twelve, and as such is naturally associated with the meaning of that number which is GOVERNMENTAL PERFECTION. It emphasizes it, and expresses spiritually the fact that perfect government of the life is only possible through the agency of the Holy Spirit. Hence it may well be called, the number of THE SPIRIT GUIDED LIFE. It occurs first in the second verse of the Bible, the Hebrew word "rachaph" translated "moved", having the number 288, or twice 144, behind it. The verse might perhaps better be given thus,

> "And the Spirit of God fluttered lovingly upon the face of the waters."

In Deuteronomy 32:11, the next place where it occurs, it is translated "fluttered", in that beautiful passage,

> "As an eagle stirreth up her nest, fluttereth over her young, spreadeth abroad her wings, taketh them, beareth them on her wings: so the Lord alone did lead him, and there was no strange god with him."

In both these passages the gracious work of the Heavenly Dove, the Holy Spirit of God, is described. In the re-Creation story, the Holy Spirit is seen moving over the dark waters of the chaos into which the earth had fallen, about to begin the work of re-storing it into an earth of order and beauty to be inhabited by man. And in Deuteronomy we see the same loving Spirit leading, watching over, and caring for the chosen nation through all their desert wanderings.

We next meet with this number at the beginning of the Eden story, where we read in Genesis 2:8,

> "And the Lord God planted a garden eastward in Eden."

The Hebrew word, "Qedem" — eastward, having the numeric value of 144, shows us the aspect of the garden, and the next few words,

> "and there He put the man whom He had formed,"

showing us God's action towards the first man He made, that it is His purpose that all His created beings should live a life of perfect government, a wonderful garden life, bringing forth the fruit of the Spirit, and finding their chief joy in communion with their loving God. How few live this life, for while many make much of the "eastward position", it is not that arising from the infilling of the Holy Spirit. As man was casted out of the east gate of the garden so man must return through the east gate. In the Tabernacle in the Wilderness God gave us one of the most beautiful pictures of communion with Him. If man is to return back into the presence of His Holy God, he must come through the east entrance, which was the only entrance to the Tabernacle. As he enters this gate, he is confronted with the blood at the Brazen Altar, then the cleansing power of God at the Laver and then into the Holy Place where daily communion is made possible as we are refilled with the Holy Spirit, the oil of the Lord, and then in prayer and feasting upon the Bread we are cleansed and sanctified to appear into the Holy of Holiest, the presence of our Lord.

Another very interesting occurrence of this number is in the cubic capacity of the Ark at the Flood. By multiplying its three dimensions of 300 cubits, 50 cubits, and 30 cubits, we find that its cubic capacity was 450,00 cubic cubits. This represents the amount of air space, Noah and his family had to live on, and we find that this figure, 450,000, divides up into 5x5x5x5x5x144! In other words, 5 to its fifth power multiplied by 144. FIVE, as we know, speaks of GRACE, so in this supply of a space of safety from the surrounding waters of judgment, we see the spiritual truth brought out that it is all of God's infinite GRACE, and under the guidance of that same Holy Spirit that fluttered lovingly over those waters at the earlier re-creation of the earth.

This truth is again symbolically expressed in Revelation, where in the seventh chapter, 144,000, are sealed by the Holy Spirit of God before God's judgments fall on earth. It does not of course mean that this identical number are saved in that time of Tribulation. Many more than this will be saved, but it is only using the numbers 144 and 10, (144,000) in the spiritual sense in which these numbers are used through-out Scripture to tell of those who by the power of the Holy Spirit are kept through that time of trouble.

We might notice here that the Greek words for one hundred and forty and four are not used but were used respectively as symbols for one hundred "r", forty "M", and four "d", as shown in the Greek alphabet given at the end of this book. One hundred and forty and four simply reading in the Greek text of Revelation 7:4, "rmd".

The same number is again used in Revelation 14, where we see 144,000 in Heaven, "having His Father's Name written on their foreheads." They are "the firstfruits unto God, and to the Lamb." Again we are given a picture of those who have yielded their lives wholly to the guidance of the Holy Spirit. They are taken up to be with Christ, their Saviour, before the actual "harvest of the earth" is reaped.

A beautiful instance of the use of this number occurs in Genesis 24:22, where Abraham's servant, a type in this chapter of the Holy Spirit, puts upon Rebekah, first, "a golden earring," or "jewell for the forehead", a sign of her conversion, and then upon her hands, "two bracelets," a sign of her consecration; these fetters are a sign of bondservice, of ownership by someone other than herself. Now the Hebrew word used for "bracelet" here is "tsamid," and its numeric value is 144. Thus symbolising that Rebekah, who is here a type of the Bride of Christ, has yielded her life wholly to her Bridegroom, and placed herself entirely under the guidance of the Holy Spirit, "a bond-servant of Jesus Christ."

Another instance of this number will be found in the face that the children of Israel remained in Egypt exactly 144 years after Joseph's death, and then the Exodus took place, and they marched out a free people under the guidance of the Pillar of Fire!

Then, when they were established years later in the Promised Land, David, under divine guidance preparing for the worship of God in the Temple to be built by Solomon, forms a choir, and we read in 1 Chronicles 25:7,

> "So the number of them, with their brethren, that were instructed in the songs of the Lord even all that were cunning, was two hundred fourscore and eight."

And this number is just twice 144! Thus typifying true Spirit-filled praise, as in Ephesians 5:18, 19.

> "Be filled with the Spirit, speaking to yourselves in psalms and hymns and spiritual songs, singing and making melody in your heart to the Lord."

Yet another instructive word "anephthe" – Kindled – occurs in Luke 12:49,

> "I am come to send fire on the earth; and what will I if it be already kindled?"

Our Lord here speaks of that baptizing with the Holy Spirit and with fire, which John the Baptist had prohesied of Him. Now the numeric value of this word, "anephthe," is 576, or FOUR times 144. FOUR is the WORLD NUMBER and it speaks of that heavenly Fire which is kindled in the soul upon the earth, when the Holy Spirit comes in to control and guides into a life of perfect government.

Many other instances occur of this number, but the last we will refer to is that of the wall of the New

Jerusalem, of which we read,

> "And he measured the wall thereof, and hundred, and forty and four cubits, according to the measure of a man, that is, of the angel." Revelation 21:17

Here we see the wall of 144 cubits in heights, surrounding the holy city, and telling us that no evil government is allowed to enter that place of God's abode. The measure is that allotted to man, the measure of the Spirit-filled life. God's command to ever believer is,

> "Be filled with the Spirit."

No evil spirit of Government can mar the believer's life and witness, if that person's will and life is controlled by the Holy Spirit of God.

ONE HUNDRED FIFTY-THREE

FRUIT-BEARING

We now come to a number that perhaps exhibits in a more striking manner than any other number, the use which the Holy Spirit makes of arithmetic in God's Word to teach deep spiritual lessons. It confirms in a marvelous way the fact of the God-breathed nature of the Word, for no human mind could possibly have devised such a means of binding the whole of the Word together, and proving beyond the shadow of a doubt in any but wilfully blind heart, that the whole of the Bible is the work of a Supreme Master Mind, the work of One Who alone "searcheth all things, yea, the deep things of God", that Divine Person Who is called "the Spirit of God".

In the Gospel of John we note that all the miracles recorded are called "signs", that is, that there is a hidden prophetic meaning lying behind each one, and when we come to the only one recorded there as performed by the Lord after His Resurrection, we find a mention in it of this special number, for in John 21:11, we read,

> "Simon Peter went up, and drew the net to land full of great fishes, and hundred and fifty and three: and for all there were so many, yet was not the net broken."

The fact that the number of fishes mentioned was not a round one, such as 150 or 200, early attracted the attention of God's saints, and it was clearly seen that there must be some special spiritual significance hidden in the careful way the exact number of fishes is detailed.

Augustine seems to have been one of the first to find the key, though even he does not seem to have seen what spiritual significance attached to it. He noticed, however, that if you put down all the numbers from one upwards in a perpendicular line so as to make an addition sum of them, and if you stop when you have put down seventeen, and then draw a line and add them up, the addition comes to exactly "ONE HUNDRED AND FIFTY-THREE". Thus, this special number is the sum of all the digits from one to that very number, SEVENTEEN, which we have seen stands for VICTORY, and is so frequently used by the Holy Spirit throughout the whole of God's Word.

When we then turn to the story, we soon see that it is a parable, acted out by the SEVEN disciples, and full of deep spiritual lessons for all the Lord's people down the ages, and especially for those living in the Time of the End, as we now are.

The disciples have been told to go into Galilee, and there wait for our Lord, but they appear to have gotten tired of doing so, and so they revert to their old occupation of fishing. They have become discouraged. They felt they could not live for the Lord if he left them since in His very presence they had failed him so often. So a crisis has come to the work of our Lord. The disciples are walking out on the Lord! They are going back to the nets, boats, and fishing. The result of a long night's toil is an empty net! Just so have we been sent into Galilee, the place of every day life, but with one sole object, "to wait for His Son from Heaven, Whom He raised from the dead, even Jesus, which delivered us from the wrath to come."

Like the disciples, we too have become tired of waiting for His "glorious appearing", and many cry in these last days,

> "Where is the promise of His Coming? for since the fathers fell asleep, all things continue as they were from the beginning of the creation." (II Peter 3:4)

The church has thus decided that it is no use waiting, and that the best thing to do is to go into the world, and try and produce a man-made Millennium. She has lost her separated character, and instead of living in heart-touch with the Master, laboring in the harvest-field to win precious souls for the Lord, and waiting patiently for His coming, she has lost her "first love", and become lukewarm. She thinks herself to be well-off and having need of nothing, and knows not that she is "wretched, and miserable, and poor, and blind, and naked." (Revelation 3:17)

Her net is empty, no souls are being won for Christ, because she is not seeking that enduement of "power from on high" without which all her labours are nothing.

Then we read,

> "when the morning was now come, Jesus stood on the shore,"

So our Saviour stands even now on the Heavenly shore, at the dawn of the glad Millennial morning, and His Word is now as then,

> "Cast the net on the right side of the ship, ye shall find."

Use your daily labour for My glory, and in accordance with My Word. We, who are redeemed by the precious Blood of Christ, are not sent into the world to earn our living, but to glorify God, to do His will. It is

> "Seek ye first the Kingdom of God and His righteousness, and all these things shall be added unto you." (Matthew 6:33)

Food and drink is not to be the object of our lives, "your Heavenly Father knoweth that ye have need of all these things."

> "For the kingdom of God is not meat and drink; but righteousness, and peace, and joy in the Holy Ghost."

The sphere of life in which God has placed us is to be used as a Gospel net to win the precious souls, with whom we come in contact, for the Saviour. Obedience to the Saviour's slightest work as it comes from the Heavenly shore is the first supreme lesson for the Lord's people to learn. Though the Saviour has left this earth, He still walks there in His sanctified people's hearts, and when His Word is obeyed, the result is a net full of fishes. Thus is the perfection of spiritual order wrought out in the world by the work of the Holy Spirit and VICTORY is assured in the hearts of believers.

But there is more in the picture than this, for as another of God's saints points out, this number 153 is 17 multiplied by 9. Now NINE is the number of THE FRUIT OF THE SPIRIT, and so this number 153 signified the VICTORY brought about IN BEARING THE FRUIT OF THE SPIRIT in our lives.

Note also the net is full, and is drawn to land.

> "and for all there were so many, yet was not the net broken."

Every single fish is brought safely ashore, none are lost,

> "those which thou gavest Me I have kept,"

says our Lord, and again,

> "no man is able to pluck them out of My Father's hand."

Thus we are given a beautiful picture of the VICTORY is obtained by the saints of God as we BEAR FRUIT for Him in this life.

TWO HUNDRED
INSUFFICIENCY

This number, TWO HUNDRED is very seldom found in the Bible but each time it is mentioned it reveals INSUFFICIENCY. TWENTY is the number of EXPECTANCY, hence we have TWENTY x TEN which equals TWO HUNDRED.

The significance of this number is suggested by John 6:7, where we read,

> "Two hundred pennyworth of bread is NOT SUFFICIENT for them."

In Joshua 7:21, we read,

> "When I saw among the spoils a goodly Babylonish garment, and TWO HUNDRED shekels of silver, and a wedge of gold of fifty shekels weight, then I coveted them, and took them; and, behold, they are hid in the earth in the midst of my tent, and the silver under it."

Here the sin of Achan is revealed and the TWO HUNDRED shekels were "NOT SUFFICIENT" to save him from the consequences of his sin. This shows us the INSUFFICIENCY OF MONEY. (Psalms 49:7-9)

Absalom's TWO HUNDRED shekels weight of hair were "NOT SUFFICIENT" to save him, but rather caused his destruction.

> "And when he polled his head, (for it was at every year's end that he polled it: because the hair was heavy on him, therefore, he polled it: he weighed the hair of his head at TWO HUNDRED shekels after the king's weight." (II Samuel 14:26)

This shows us the INSUFFICIENCY of beauty. (II Samuel 18:9)

Micah's graven image was purchased for TWO HUNDRED shekels and led to the introduction of idolatry into Israel and the blotting out of the tribes of Dan and Ephraim from the blessings of God as revealed in Revelation 7. This shows us the INSUFFICIENCY of mere religion.

> "Yet he restored the money unto his mother; and his mother took TWO HUNDRED shekels of silver, and gave them to the founder, who made thereof a graven image and a molten image: and they were in the house of Micah." (Judges 17:4)

This number is mentioned only six times in the Bible.

SIX HUNDRED
WARFARE

SIX HUNDRED is the number that is connected with WARFARE. Pharoah pursued Israel with SIX HUNDRED chariots.

> "And he took SIX HUNDRED chosen chariots, and all the chariots of Egypt, and captains over every one of them." (Exodus 14:7)

Israel, under Shamgar, slew SIX HUNDRED of the Philistines with ox goads. (Judges 3:31)

SIX HUNDRED of the soldiers of the tribe of Benjamin escaped slaughter in a battle and hid in the rock of Rimmon. (Judges 20:46-47)

The Danities sent up SIX HUNDRED men with weapons of war to take the city of Laish. (Judges 18: 7-11)

Goliath's spear weighted SIX HUNDRED shekels of iron. (I Samuel 17:7)

David had about SIX HUNDRED men with him when he was being pursued by Saul. (I Kings 23:13)

With SIX HUNDRED men David pursued and defeated the Amalekites. (I Samuel 30:1-18)

SIX HUNDRED SIXTY-SIX

THE NUMBER OF THE BEAST

ANTI-CHRIST

This remarkable number has for many centuries been the subject of comment, because of its prominence in the thirteenth chapter of Revelation, where we are definitely invited to solve its meaning. In verse 18 we read:

> "Here is WISDOM. Let him that hath understanding COUNT the number of the BEAST: for it is the number of a man; and his number is SIX HUNDRED THREE-SCORE AND SIX." (Revelation 13:18)

Now that a knowledge of the meaning of the various numbers has been acquired the NUMBER OF THE BEAST can be COUNTED, and the Bible shows just where to start counting. Revelation 13:18 which gives the number of the Beast, and tells the reader to count that number, starts by saying, "Here is WISDOM". Revelation 17:9 opens by saying,

> "And here is the mind WHICH HATH WIDSOM."

That statement is followed by a series of numbers which add up to 666, the number of the BEAST. The passage must be read and counted to Revelation 17:14. That verse says,

> "THESE shall make WAR with the Lamb, and the Lamb shall overcome them."

The number for WARFARE is 600.

In VERSE 9 there are 7 heads, 7 mountains, and 1 woman.

In VERSE 10 there are 7 kings; 5 are fallen, 1 that is, and 1 that is not yet come.

In VERSE 11 it is said that the beast was the 8th, and was of the 7th, and that he goeth into perdition.

In VERSE 12 there are 10 horns, and 10 kings, and 1 hour.

In VERSE 13 they have 1 mind, and give their strength and power to the beast.

In VERSE 14 there is war, and the number for that is 600.

Add up these numbers. The total is 666. It is amazing! These numbers are all taken from the Bible itself.

If the reader will refer back to the numbers that add up to 666, he will find that the 5th number in the list is FIVE, the 8th number is EIGHT, and the 10th number is TEN. John said,

> "Let him that hath understanding, COUNT the NUMBER of the Beast."

The numbers 5, 8, and 10 are the numbers that coincide with themselves in the list. These three numbers when added together, equal 23, or DEATH. Those who refuse to submit to the authority of the beast will be killed, unless they go into hiding.

The BEAST'S NUMBER can also be reached by adding 66, the number for idol worship, and 600, the number for warfare. Both idol worship and warfare will be connected with the beast. I ask you to read Revelation 13:13-15.

There were three men in Scripture which stand out as the avowed enemy of God and His people. Each is branded with the number SIX that we may not miss their significance.

1. GOLIATH, whose height was 6 cubits, and he had 6 pieces of armour; his spear's head weighed 600 shekels of iron. (I Samuel 17:4-7)
2. NEBUCHADNEZZAR, whose "image" which he set up, was 60 cubits broad (Daniel

3:1), and which was worshipped when the music was heard from 6 specified instruments.

3. ANTICHRIST, whose number is 666 as revealed in Revelation 13:18.

In the first we have ONE SIX connected with the PRIDE of FLESHLY might.
In the second we have TWO SIXES connected with the PRIDE of absolute DOMINION.
In the third we have THREE SIXES connected with the PRIDE of SATANIC guidance.

Now many thousands of the Greek words, of which the New Testament is composed, have been counted, and only FIVE words have so far been discovered which have the number 666 hidden behind them. The first of these FIVE words occurs in the story of the sudden storm on the lake of Galilee. The Lord Jesus is asleep in the boat, and terrified by the violence of the storm, the disciples awake Him with the cry, "Lord, save us: we perish." The incident is related in Matthew, Mark and Luke, and through the wording of the cry is different in each case, the Holy Spirit evidently giving the despairing cry of three different disciples in the boat, yet the same word, "apollumetha" — "we perish", occurs in each Gospel, and the number behind it is the mystic number 666!

Like so many of our Lord's miracles, the incident is evidently a typical one, and shows us the picture of the Church of Christ in jeopardy. The cause of the danger is described by each writer, it is that the boat was becoming filled with water. Now water is a type of the world, the church is intended to be in the world, even as the ship is in the sea, but just as it is dangerous to let the water get into the ship, so it is dangerous when the world gets into the church.

> "They are not of the world, even as I am not of the world." (John 17:16)

says our Lord to His disciples. The world, however, got into the church in its early days, and now the church is nearly full of the world, even as "the ship was covered with the waves". Dances, theatricals, social hours, worldly singing are common things in churches, and it is all done in the name of religion!

Thus we have reached the Time of the End, when the true Church of Christ, imperiled by the near setting-up of the Anti-Christ, is crying to the Saviour to save it from the danger, and according to His Promise, call it up to meet Him in the air!

Hence the first use of this number is in a type picture of the deadly peril of the church, when the world has filled it, and its rescue from this danger by the voice of Christ, Himself stilling the storm.

The second word with this number is "PARADOSIS" -- "TRADITION". It occurs significantly THIRTEEN times only! The first place is in Matthew 15:2, 3, 6, where our Lord ask the Pharisees,

> "Why do ye also transgress the commandments of God by your tradition?"

He adds, quoting from Isaiah,

> "But in vain they do worship Me, teaching for doctrines the commandments of men."

The chief peril of the true church consists of departure from God's own Word. The largest nominal body, the Roman Catholic Church, deliberately places the teaching of "Mother Church" above the "God-breathed" Scriptures. This destructive characteristic of the nominal church is stamped by God, through this word "tradition", with this terrible number 666. The chart and compass of God's people is ever God's Holy Word. No traditions of men, mis-called church teaching, should ever be allowed to supplant it.

The third word is "PLEURAN" -- "SIDE", and occurs FIVE times, first in John 19:34, where we read,

> "But one of the soldiers with a spear pierced His side, and forthwith came there out blood and water."

This word "pleura" is always used in the accusative case, "pleuran", and in this form it counts up to 666! (Also read John 20:20, 25, 27)

Our Lord was already dead. The Atoning Sacrifice was complete, there needed nothing further, the work was done when He cried, "It is finished" and "gave up the ghost". The offering had been made "once for all", it needs no repetition. Yet here was one, a Roman, daring to pretend to finish the work, and today he has many imitators, who offers what they call the sacrifice of the mass, which they declare to be an actual-re-enactment of the dread scene on Calvary. All such "blasphemous fables", as our Book of Common Prayer calls the Sacrifices of Masses, God stamps with the awful number 666, the number of the Beast!

The fourth word is "EUPORIA" -- "WEALTH". It only occurs in Acts 19:25. Demetrius, a silversmith, who made silver shrines for Diana, thereby getting "no small gain", calls together his fellow-craftsmen, and says to them,

> "Sir, ye know that by this craft we have our wealth."

This word "wealth", then, is stamped by God with this dreadful number 666, for the worship of graven images is an abomination in God's sight. Yet this terrible sin is rampant in the nominal church today, and houses of prayer are filled with images, while many are erected in the streets, and thousands set up graven images in their own homes before which they bow. It is all leading up rapidly to that day when the Image of the Beast, whose number is 666, will be set up, and worship of it will be compulsory on pain of death. That terrible man of sin, who will declare himself to be God incarnate!

The fifth word will be found in I Peter 1:1, "DIASPORAS" -- "SCATTERED". This letter is thus addressed to the "scattered" saints of God, and this word is stamped with this number 666, because the true Church of God would be persecuted and "scattered" down all the centuries by those who loved the world, traditions, the sacrifice of the mass, wealth, and image worship, better than the things of God. The Beast, which has scattered the true church, is rearing today more arrogantly than ever, and conditions are rapidly maturing for the final full revelation of the Mystery of Iniquity, following on the Rapture of the Saints, whose "number", as we shall see in the next chapter, 888, is a very different one from that of the Beast, whose "number is six hundred, threescore and six."

EIGHT HUNDRED EIGHTY-EIGHT

THE FIRST RESURRECTION SAINTS

In the previous number, 666, we have man's number, six, in a triple form, shewing how man, rebelling against God, would eventually accept the overlordship of a super man, the Man of Sin, Anti-Christ, whose awful reign will commence at the close of this present age.

Now we shall consider another number in a triple form, the number of the FIRST RESURRECTION OF THE SAINTS, number 888. In this number we shall see how man, when he accepts God's will as his rule of life is kept safe in the hour of danger, and how the coming of the only Perfect Man who has ever willingly accepted His overlordship, and a glad meeting with Him at the appointed meeting place in the air.

The starting place where we shall meet with this number 888, is in the early chapters of the Bible. Man has so signally failed in the purpose for which he was created, that of walking like Enoch in holy fellowship with his maker, that the flood is the only remedy to stem the ever-increasing tide of human wickedness. But God in His infinite mercy provides an ark of safety for all who will listen to His voice from the coming judgment. Of all the human beings populating the earth, Noah and his family, numbering eight person, alone heed the warning, and accepting God's loving invitation are sheltered safely in the ark while the waters cover the earth. We know that names of the four men who thus found safety -- Noah, Shem, Ham and Japheth. Of these four, Ham sinned after coming out of the ark, and was cursed for mocking his father, Noah. Thus the others who were not under a curse, the good men, who believed God's word are Noah, Shem, and Japheth.

Noah	. . . . Numerics Value	58
Shem		340
Japheth		490
		888

Thus in the numeric value of these names added together gives us a picture of the resurrection, the first fruits, of these three who did not come under judgment.

Many centuries later, the Israelites whom God had chosen out to be an elect nation of witness, had forsaken the path God had marked out for them, and intent upon their own lustful desires, had fallen into idolatry. God consequently visited them with judgment, just as He had visited the antediluvian race, and allowed them to be overcome by their enemies, and finally carried away to Babylon. But as at the flood, even in this time of evil, He yet had a faithful few who, like Noah, Shem, and Japheth, were true to the God they trusted and loved. And in the book of Daniel we read in the first chapter of four men who stood firmly out for God, daring the king's wrath, and facing the SEVEN times heated furnace of fire, and the lion's den, in believing faith that their God was able to deliver them in the hour of trial.

These FOUR men bear the names of Daniel, Hananiah, Mishael, and Azariah. And when we add together the numeric value of these names, we again find the total to be the number we are studying, 888!

Daniel	95
Hananiah	120
Mishael	381
Azariah	292
	888

Here again God was faithful to those who trusted in His Word, and as at the flood, Noah, Shem, and Japheth were saved from judgment, so these four young princes of Judah proved their God to be a God "able to deliver from judgment". And so they received a joyful resurrection, as it were, from what seemed certain destruction, neither the burning flames, nor the lions' mouths being able to harm those who look wholly to God for deliverance.

Again, centuries pass, Babylon, Medo-Persia, and Greece, great empires though they were, have crashed into ruins and another empire now holds sway over all the earth, Rome, mightiest of them all! Yet still the great mathematician, called in Daniel 8:13, "as we have seen, "the wonderful numberer," orders all things according to His unfettered Will. And so Rome issues a decree that all the world shall be enrolled in a census, and this decree sends Mary and Joseph to the little town of Bethlehem, where God has promised the long-looked-for Saviour, and there a baby is born! And before His birth Joseph is told he is to call the baby by the name, Jesus, "for He shall save His people from their sins."

This is the promised Saviour to whom Noah, and his sons at the flood, and Daniel and his comrades in Babylon, all looked forward in faith, and when we count the number of His sacred name we find again the same number that was hidden behind the names of those faithful disciples so many centuries before! We give it in full:

I	----------	10
E	----------	8
S	----------	200
O	----------	70
U	----------	400
S	----------	**200**
		888

Salvation from the flood of waters, from the burning fiery furnace, and from the lions' den, is all through faith in Him who is "the resurrection and the life." the number of whose name is 888.

Thus from Genesis, right onward through the Bible, we see God using this beautiful number 888 to confirm the truth of His Word. It leads up from the flood right to the rapture, that rapture which is very near, and may occur at any moment now.

THOUSAND
DIVINE COMPLETENESS AND THE GLORY OF GOD

This number is connected with DIVINE COMPLETENESS AND THE GLORY OF GOD. It is usually used with angels especially in the book of Revelation. It reveals His Divine care and protection to those who are in Christ Jesus.

The highest number referred to in any way in the Bible is ten thousands times ten thousand;

> "A fiery stream issued and came forth from before him: thousand thousands ministered unto him, and ten thousand times ten thousands stood before him: the judgment was set, and the books were opened." (Daniel 7:10)

> "And I beheld, and I heard the voice of many angels round about the throne and the beasts and the elders: and the number of them was ten thousand times ten thousand, and thousands of thousands;" (Revelation 5:11)

The Hebrews had no number for million so they used thousand thousand to represent a million. Always we see the GLORY OF GOD in the number THOUSAND.

GEMATRIA

Gematria what the ancients called the use of the Greek and Hebrew alphabets, when they used them to represent their numbers instead of using the Arabic figures as we do. Note the chart of the two alphabets below. Each letter has its numerical valuation. This will work only in the Greek and Hebrew.

THE HEBREW ALPHABET

consists of 22 (2 x 11) letters, so the 5 finals were added to make up three series of 9, or 27 in all, but are not used in Gematria.

Aleph	א	= 1	Yod	י	= 10	Koph	ק	= 100	
Beth	ב	= 2	Kaph	כ	= 20	Resh	ר	= 200	
Gimel	ג	= 3	Lamed	ל	= 30	Shin	ש	= 300	
Daleth	ד	= 4	Mem	מ	= 40	Tau	ת	= 400	
He	ה	= 5	Nun	נ	= 50	Kaph	ך	= 500	Finals.
Vau	ו	= 6	Samech	ס	= 60	Mem	ם	= 600	Finals.
Zayin	ז	= 7	Ayin	ע	= 70	Nun	ן	= 700	Finals.
Cheth	ח	= 8	Pe	פ	= 80	Pe	ף	= 800	Finals.
Teth	ט	= 9	Tsaddi	צ	= 90	Tsaddi	ץ	= 900	Finals.

THE GREEK ALPHABET

The Greek letters were 24, and the required number, 27, was made up by using the final "s" or s (called Stigma) for 6, and adding two arbitrary symbols called Koppa, for 90, and Sampsi, for 900.

Alpha	α	= 1	Iota	ι	= 10	Rho	ρ	= 100
Beta	β	= 2	Kappa	κ	= 20	Sigma	σ	= 200
Gamma	γ	= 3	Lambda	λ	= 30	Tau	τ	= 300
Delta	δ	= 4	Mu	μ	= 40	Upsilon	υ	= 400
Epsilon	ε	= 5	Nu	ν	= 50	Phi	φ	= 500
Stigma	ς°	= 6	Xi	ξ	= 60	Chi	χ	= 600
Zeta	ζ	= 7	Omicron	ο	= 70	Psi	ψ	= 700
Eta	η	= 8	Pi	π	= 80	Omega	ω	= 800
Theta	θ	= 9	*Koppa*	Ϙ	= 90	*Sampsi*	ϡ	= 900

*This letter ϛ (called Stigma) is used for the number 6. Why this letter and number should be thus associated we cannot tell, except that both are intimately connected with the ancient Egyptian "mysteries". The three letters SSS (in Greek ϛϛϛ) were the symbol of Isis, which is thus connected with 666. Indeed the expression of this number, χξϛ, consists of the initial and final letters of the word Χριστός (Christos), Christ vix., X and s, with the symbol of the serpent between them, χ -ξ- ϛ

VALUES OF THE HEBREW AND GREEK ALPHABETS

To complete our study on the numbers of the Bible it would be well for us to consider the numeric value system of the Hebrew and Greek alphabets. When you and I wish you say "1" we do not spell "one" but we write the figure "1", and the same for "2" and all the other figures. In other words we have distinct symbols for numbers. We use the Arabic figures 1, 2, 3, 4, 5, 6, 7, 8, 9, 0. But the Greeks in whose language the New Testament was written, and the Hebrews, in whose language the Old Testament was written, had no special symbols for numbers. When they wished to say "1" they wrote the first letter of the alphabet. They had a definite system where every letter of the Greek language stood for a special number. The first nine letters stood for 1-10, then from 20-100, and from 100-400. The Hebrews had the same system. They had 22 letters instead of 24, but each letter stood for a number, and each word is simply a sum in arithmetic, by adding the numeric values of the special letters.

As each word consists of letters THE NUMERIC VALUE OF A WORD IS THE SUM OF THE NUMERIC VALUES OF ITS LETTERS. The numeric values of the words of which these consist.

By means of these numeric values the Greek and Hebrews performed all their numeric operations. But in scripture an additional system is made use of for the purpose of numeric construction of the text, that of PLACE VALUES.

The PLACE VALUE of a letter in Scripture, whether Hebrew or Greek is the number of the PLACE the letter occupies in the alphabet. Accordingly in the Hebrew the place values and the numeric values of the first ten letters are the same, and the same is the case with the first five letters in the Greek. But the ELEVENTH Hebrew letter does not stand for eleven, but twenty. Accordingly its numeric value is 20, but its place value is 11; the last letter of the Hebrew alphabet, the twenty second, stands for 400. Accordingly its numeric value is 400, but its place value is 22. The same applies to the Greek alphabet. Its sixth letter stands for 7; this is its numeric value, but its place value is 6.

The VALUE of a Hebrew or Greek letter or word is the SUM OF ITS NUMERIC AND PLACE VALUES: thus the VALUE OF JESUS, in GREEK is 975, of which the NUMERIC VALUE IS 888, and the place value is 87.

The numeric value of the Name of our Lord -- "JESUS" - in Greek is 888. I have chosen that name because it is a good illustration. We are living in times when the anti-Christ is to be revealed soon, and the Scriptures warn us -- "who hath wisdom let him count, for his name is the name of a man." And the number is 666 and our Lord's number is 888. Anyone a little familiar with spiritual things would know at once that this collocation of 666 and 888 is not accidental; so we can see the Trinity of the Evil One and the Trinity of the New Creation.

There is not a paragraph in the Bible that is not constructed on a similar numerical scheme. Even if men had the skill to construct the 66 books of the Bible, the construction of even a single book would in nearly all cases require some hundreds of years. As no man or set of men could write books thus, the only rational explanation of the numeric phenomena of the Bible is the guidance of its writers by a superior mathematical mind, the mathematical Author of Nature. These phenomena thus prove the Verbal Inspiration of the Scriptures as a whole and in every part thereof, even to its letters, in the original language.

The author has not made a thorough study of the NUMERICAL VALUE OR PLACE VALUE of the Scriptures but is just beginning on this phase of number study. A whole life time could be devoted to such a study and not get out of one of the 66 books of God's Word. The reader is challenged to get a Greek Lexicon and begin now to COUNT the letters and words in the Word of God.

Finally, in these closing words, the writer wishes

to exhort any who may chance to read these pages, if he has not already done so, to prepare to meet Him who is the great Author of the BOOK of books, the Bible. Turn away from the wisdom of men. Cease to depend on the wisdom and works of the flesh, and turn to Him who is higher than the heavens, and whose wisdom and power can never be measured. Repent of your sins and place your trust in Christ who died on the cross to redeem us from sin, and who arose again three days later and then ascended to heaven to sit at the right hand of the Father. Right now ask Him to forgive you and save you.

> "For whosoever shall call upon the name of the Lord shall be saved." (Romans 10:13)

Now,

> "Unto him that loved us, and washed us from our sins in his own blood, and hath made us kings and priests unto God and his Father; to him be glory and dominion for ever and ever. AMEN"

If you will trust Christ and take him as your own Saviour, how glad I would be to hear from you. If you can do so honestly, I beg you to ask Christ to save you and accept Him as your salvation. Then write me at the address below. I will send you a letter of counsel and encouragement and will rejoice with you.

EVANGELIST ED. F. VALLOWE
P.O. Box 826
Forest Park, GA 30098–0826

BIBLIOGRAPHY

Bullinger, J. W. — Numbers of the Scriptures, London, England - 1926

Jones, G. E. — That Ye may Marvel, Sammons Printing Company, Jonesboro, Arkansas - 1953

Kesner, J. W. — Numbers In The Scriptures, J. W. Kesner & Company, Ft. Smith, Arkansas - 1955

Larkin, Clarence — Dispensational Truth, Clarence Larkin Est., Philadelphia 32, Pa. - 1920

Encyclopedia Britannica, Volume 14 — Encyclopedia Britannica, Inc., William Benton, Publisher, Chicago, Illinois - 1962

Gaebelein, Frank E. — Exploring The Bible, Van Kempen Press, Wheaton, Illinois - 1950

Lamsa, George M. — New Testament Commentary, A. J. Holman Co. - 1945

Harrison, Sale L. — The Wonders Of The Great Unveiling, Evangelical Press, Harrisburg, Pa. - 1930

De Haan, M. R. — Revelation, Zondervan Publishing Co., Grand Rapids, Mich. - 1946

Naish, Reginald T. — Spiritual Arithmetic - 1926, C. J. Thynne & Jarvis Ltd.

Panin, Ivan — Shorter Works Of Ivan Panin, The British Israel Assoc., London, England - 1934

Panin, Ivan — Bible Numerics, The British Israel Assoc., London, England - 1934

Summers, Ray — Worthy Is The Lamb, The Broadman Press, Nashville, Tennessee - 1950

Walvoord, John F. — The Millennial Kingdom, Dunham Publishing Co., Findlay, Ohio - 1959